Practical Pharmacology for the Pharmaceutical Sciences

Practical Pharmacology for the Pharmaceutical Sciences

D. Michael Salmon

School of Health and Biosciences, University of East London, UK

This edition first published 2014
© 2014 John Wiley & Sons, Ltd

Registered office
John Wiley & Sons Ltd, The Atrium, Southern Gate, Chichester, West Sussex, PO19 8SQ,
United Kingdom

For details of our global editorial offices, for customer services and for information about how to apply for permission to reuse the copyright material in this book please see our website at www.wiley.com.

Wiley also publishes its books in a variety of electronic formats. Some content that appears in print may not be available in electronic books.

Designations used by companies to distinguish their products are often claimed as trademarks. All brand names and product names used in this book are trade names, service marks, trademarks or registered trademarks of their respective owners. The publisher is not associated with any product or vendor mentioned in this book.

Limit of Liability/Disclaimer of Warranty: While the publisher and author have used their best efforts in preparing this book, they make no representations or warranties with respect to the accuracy or completeness of the contents of this book and specifically disclaim any implied warranties of merchantability or fitness for a particular purpose. It is sold on the understanding that the publisher is not engaged in rendering professional services and neither the publisher nor the author shall be liable for damages arising herefrom. If professional advice or other expert assistance is required, the services of a competent professional should be sought.

The advice and strategies contained herein may not be suitable for every situation. In view of ongoing research, equipment modifications, changes in governmental regulations, and the constant flow of information relating to the use of experimental reagents, equipment, and devices, the reader is urged to review and evaluate the information provided in the package insert or instructions for each chemical, piece of equipment, reagent, or device for, among other things, any changes in the instructions or indication of usage and for added warnings and precautions. The fact that an organization or Website is referred to in this work as a citation and/or a potential source of further information does not mean that the author or the publisher endorses the information the organization or Website may provide or recommendations it may make. Further, readers should be aware that Internet Websites listed in this work may have changed or disappeared between when this work was written and when it is read. No warranty may be created or extended by any promotional statements for this work. Neither the publisher nor the author shall be liable for any damages arising herefrom.

Library of Congress Cataloging-in-Publication Data

Salmon, Michael (D. Michael), author.
  Practical pharmacology for the pharmaceutical sciences / Dr. Michael Salmon.
    p. ; cm.
  Includes bibliographical references and index.
  ISBN 978-1-119-97550-2 (pbk.)
  I. Title.
  [DNLM: 1. Pharmacology–Laboratory Manuals.   2. Chemistry, Pharmaceutical–Laboratory Manuals.   3. Pharmaceutical Preparations–Laboratory Manuals.   QV 25]
  RM301.25
  615.1078–dc23

                                                                                    2013037819

A catalogue record for this book is available from the British Library.

ISBN: 9781119975502

Set in 10.5/13pt Sabon by Aptara Inc., New Delhi, India

Printed and bound in Malaysia by Vivar Printing Sdn Bhd

1   2014

Contents

Preface   ix

Acknowledgements   xi

1   Before Entering the Pharmacology Laboratory   1
    1.1   Safety and Risk Assessment   1
    1.2   The Laboratory Record Book   3
    1.3   Use of Animals in Practical Pharmacology   4
    1.4   Experimental Design   5
    1.5   Units, Dilutions and Logarithms   7
          1.5.1   Units of Mass   8
          1.5.2   Units, Concentrations and Logarithms   8
          1.5.3   Dilutions   9
          1.5.4   Logarithms   10
    1.6   Essential Statistics   12
          1.6.1   Continuous Data – t-test, ANOVA,
                  Non-parametric Tests and Regression   12
          1.6.2   Discontinuous Data – χ^2 and Fisher's Exact Test   21
    References   25

2   Basic Pharmacological Principles   27
    2.1   Drug–Receptor Interaction   27
          2.1.1   Agonists   27
          2.1.2   Antagonists   30
          2.1.3   Receptor Classification   36
    2.2   Bioassays   37
          2.2.1   Single-point Assays   38

2.2.2 Bracketing Assays, Three-point or 2×1 Assays   38
2.2.3 Multi-point Assays, Such As Four-point or
        2×2 Assays   39
References   40

3 **Isolated Tissues and Organs**   **43**
3.1 Equipment for *In Vitro* Experiments   44
3.2 Organ Baths   45
3.3 Physiological Salt Solutions   46
3.4 Transducers   47
3.5 Recording Equipment and Software   49
3.6 Dosing   50
3.7 Electrically Stimulated Preparations   52
3.8 Fault-Finding of *In Vitro* Isolated Tissue Preparations   53
References   54

4 **Smooth Muscle Preparations**   **55**
4.1 Gastrointestinal Smooth Muscle Preparations   55
4.2 Guinea Pig Isolated Ileum   56
      4.2.1 Concentration–Response Curves
              for Cholinesters   57
      4.2.2 Selective Antagonism   59
      4.2.3 Specificity of Blood Cholinesterases   62
      4.2.4 Quantification of the Potency of an Antagonist   64
      4.2.5 Bioassays   67
      4.2.6 Calcium Channel Blockers   73
      4.2.7 Field-stimulated Guinea Pig Isolated Ileum   76
4.3 Rabbit Isolated Jejunum and the Finkleman Preparation   78
      4.3.1 Adrenoceptor Sub-types   79
4.4 Isolated Tracheal Rings   80
4.5 Isolated Vas Deferens   82
Questions on Isolated Tissue Preparations   83
Answers to Problems   87
References   92

5 **Cardiovascular Preparations**   **93**
5.1 Isolated Perfused Heart Preparations   94
      5.1.1 The Langendorff Preparation   95
      5.1.2 Cardiac Interactions of Anti-asthma Drugs   98
      5.1.3 The Rat Isolated Auricle Preparation   99

5.2   Thoracic Aorta Preparation   102
      5.2.1   Drugs Regulating Nitric Oxide-mediated
              Relaxation   104
References   105

6   Skeletal Muscle   107
    6.1   Types of Skeletal Muscle   107
    6.2   Multiply-Innervated Skeletal Muscle Preparations   108
          6.2.1   Agonists and Antagonists Acting on the Frog
                  Rectus Abdominis   109
          6.2.2   Action of Anticholinesterases on the Dorsal
                  Muscle of the Leech   111
    6.3   Focally Innervated Skeletal Muscle Preparations   116
          6.3.1   The Frog Gastrocnemius Muscle–Sciatic
                  Nerve Preparation   119
    References   120

7   Isolated Cells   121
    7.1   Freshly Isolated and Cultured Cells   121
          7.1.1   Advantages of Isolated Cells   121
          7.1.2   Cultured Cells   122
          7.1.3   Cell Counting   122
    7.2   Platelets   125
          7.2.1   Inhibition of Aggregation by Nitric Oxide
                  Donors   127
    7.3   Neutrophils   131
          7.3.1   Measurement of NADPH Cytochrome c
                  Reductase   132
          7.3.2   Measurement of Intracellular $[Ca^{2+}]$ 134
    References   139

8   Biochemical Pharmacology   141
    8.1   Pharmacological Applications of Common Biochemical
          Techniques   141
    8.2   Enzyme Inhibitors   142
    8.3   Acetylcholinesterase Inhibitors   143
    8.4   Monoamine Oxidase Inhibitors   145
          8.4.1   Sub cellular Distribution of MAO Activity   146
          8.4.2   Specificity of MAO Inhibitors for Isoenzymes   149
    8.5   Thrombin Inhibitors   151
    8.6   ATPase Inhibitors   155
    References   158

9   Complementary Methods for Teaching Practical
    Pharmacology   161
    9.1   The Comparative Merits of Available Methods   161
    9.2   Interpretation of Experimental Data   162
           9.2.1   Behavioural Experiments   162
           9.2.2   Analysis of Metabolites of
                   5-hydroxytryptamine   166
           9.2.3   Radioligand Binding   167
    Answers to Questions   171
    References   176

10 Communicating Results   177
    10.1   Preliminary Reports   177
           10.1.1   Tables   178
           10.1.2   Graphs   178
           10.1.3   Bar Graphs   179
           10.1.4   Preliminary Conclusions   179
    10.2   Poster Presentations   180
    10.3   Oral Presentations   181
    10.4   Project Reports   183
    10.5   Pharmacological Literature   184
    10.6   How to Cite Scientific Information Sources   187
    10.7   Plagiarism   188
    References   188

Appendix 1: Molecular Weights of Commonly Used Drugs   189

Appendix 2: Useful Resources for Practical Pharmacology   191

Index   193

Preface

It is a startling fact that it is 30 years since the last text book on pharmacology laboratory practicals (Kitchen, 1984) was published. An obvious assumption would be that there has been a drastic fall in demand. A common response is that laboratory practicals are redundant and have been replaced by computer-assisted learning (CAL) and simulated experiments (Hughes, 2003). This is due to the challenges of increasing student numbers, decline in staff numbers and the high cost of maintaining laboratories and animal facilities (Hughes, 2001). It has been claimed that CAL experiments and problem-based learning provide an equal or superior student learning experience (Hughes, 2001, 2002). Yet surveys of the curriculum of pharmacology courses in the United Kingdom (Dewhurst and Page, 1998), and currently using a world-wide Internet survey, quickly reveals that this is not the case. "Wet" laboratory practicals remain a central part of most courses, and clearly many universities and colleges are reluctant to abandon them completely, as employers in the pharmaceutical industry and academia expect hands-on experience of pharmacological techniques. In the United Kingdom, the British Pharmacology Society (www.bps.ac.uk/Education/University resources/Core curricula in pharmacology, (accessed June 2013)) currently recommends an undergraduate pharmacology core curriculum in which skills in pharmacological experimentation form an essential component (see www.bps.ac.uk/education/university resources/core curricula). Most courses now appear to rely on in-house schedules of variable quality. Meanwhile, the equipment available for the pharmacology laboratory has greatly improved mainly through the use of computers to control experiments and record data, which makes them easier to use and improve the quality of data obtained by novice students. It is a fear that some of the skills involved in real, wet experimentation may be lost as new lecturers themselves have not been taught these methods. It is therefore timely to produce a book for use in pharmacology practical classes using state-of-the-art equipment and using modern nomenclature. Several books have recently described laboratory techniques and calculations for the biosciences in general. In contrast, this book specifically aims to introduce practical pharmacology to the pharmaceutical sciences undergraduate student.

The book opens with an outline of how to prepare to work in the pharmacology laboratory, and progresses to briefly describe some of the basic principles of pharmacology, which I believe are most clearly understood from a historical perspective. The central core includes experiments using *in vitro* tissues, isolated cells and cell-free biochemical systems, focusing on those that are unique to pharmacology. Some of these are classical experiments which were introduced some years ago, and form the basis of the discipline of pharmacology. However, it is important to note that they are still topical in that they are still being interpreted in new ways in the light of current research. Several techniques included in the BPS core curriculum, such as molecular biology, biochemistry, electrophysiology and tissue culture widely used in pharmacological research are only alluded to, as these have been well covered elsewhere. In conclusion, since no experiment is complete without communicating the results, there is a section on the presentation and interpretation of results and how to use and cite information sources. This book aspires to be useful for students in all pharmaceutical science courses that include pharmacology modules giving a real life experience in learning pharmacology.

Powerpoint slides to accompany this book can be downloaded from http://booksupport.wiley.com by entering the book name, author or isbn information.

REFERENCES

Dewhurst, D.G. and Page, C.P. (1998) A survey of the content of B.Sc. courses in Pharmacology in UK universities – is it time for a core curriculum? *Trends Pharmacol. Sci.* 19: 262–265.

Hughes, I.E. (2001) Do computer simulations of laboratory practicals meet learning needs? *Trends Pharmacol. Sci.* 22: 71–74.

Hughes, I.E. (2002) Simulations – as much learning as with wet practicals? in *Teaching Pharmacology Tomorrow: Tools and Techniques.*

Hughes, I.E. (2003) Teaching pharmacology in 2010–new knowledge, new tools, new attitudes. *Folia. Pharmacol JPN* 122: 411–418. doi:10.1254/fpj.122.41

Kitchen, I. (1984) *Textbook of In Vitro Practical Pharmacology*, Blackwell Scientific Publications. ISBN 0-632-01216-1.

Acknowledgements

I am indebted to many colleagues at the University of East London with whom I worked for many years. Without their experience and fortitude this book would not have been written. I am especially indebted to my fellow lecturers, Barry Jones, Alun Morrinan, Wilson Steele, Pat Freeman and Gill Sturman. I also received unstinting technical support from Nick Seeley and Kevin Clough. Finally, I must take the blame for any errors in this book that have been overlooked, and would be most grateful if they are brought to my attention.

1

Before Entering the Pharmacology Laboratory

Before embarking on any new activity, it is wise to be familiar with the language, concepts and possible risks of the venture. So this book begins with a number of topics with which an experimenter must be familiar, such as health and safety, ethical and legal considerations and fundamental principles of experimental pharmacology. No experiment has much value unless a coherent design has been devised first. The design of an experiment is crucial if it is to yield meaningful results. Having obtained the experimental data, it is important to decide on the relevant statistical methods that will be employed to evaluate the results. Obvious as this may seem, it is shocking, even in professional research, how many experiments are wasted due to a lack of planning in design.

1.1 SAFETY AND RISK ASSESSMENT

All activities which involve the use of chemicals, from the factory floor to the research laboratory, are subject to the Health and Safety legislation. In the United Kingdom, this is done by the Health and Safety Executive (HSE), and of particular relevance in the laboratory is the Control of Substances Hazardous to Health Regulations (COSSH, 2002). In the United States, the body is the Occupational Safety and Health Administration (OSHA), who require a Chemical Hygiene Plan (CHP) for each

Practical Pharmacology for the Pharmaceutical Sciences, First Edition. D. Michael Salmon.
© 2014 John Wiley & Sons, Ltd. Published 2014 by John Wiley & Sons, Ltd.

experiment, whilst in the EC the relevant body is the European Agency for Safety and Health at Work (EU-OSHA).

In the United Kingdom, COSSH regulations apply to all places of work, and all workers must be conversant with all risks and safety procedures. A risk assessment of all procedures must be carried out and a documentation of how these risks are to be minimized during the procedure and safe procedures for disposal of chemicals must be displayed. Any accidents must be reported and logged for future reference.

The bioscience laboratory presents many hazards not encountered elsewhere, and COSSH regulations are especially important. All laboratory workers must be aware of the regulations governing all work in laboratories. Drinking, eating and smoking are banned in all laboratories. No chemicals should come into contact with the body – including the mouth, eyes and skin whilst inside a laboratory. Remember that in a pharmacology laboratory, there is exposure to many highly biologically potent chemicals. A protective coat (frequently white) must be worn at all times, and protective eye goggles and gloves worn when required. In addition, laboratory workers must be familiar with the international warning symbols for toxic, corrosive and inflammable chemicals and gases, cancer-causing and suspected cancer-causing chemicals, radioactive materials, biological hazards, and reproductive hazards. These are widely available and explained on the internet. If a student is unsure of the meaning of any symbols they should ask their supervisor. These are not only displayed at the entrance to laboratories, but also on individual chemicals and equipment.

For class laboratory exercises, it is the responsibility of the supervisor to identify all risks and display them in the laboratory. This is not just a piece of administration, but an important document all students must be familiar with before they start the experiment to ensure safe practice. Students must be familiar with all the chemicals to be used and if there are any special precautions that must be taken. Before a project is undertaken, a risk assessment must be carried out by the student with appropriate guidance. The information that must be sought is given as follows.

- What are the dangers of handling individual chemicals? These are shown on data sheets supplied by chemical distributors. It should be ascertained if there are any particular hazards associated with entry into the body of any of the chemicals; are any substances absorbed by the skin or inhaled through the nose? Precautions that might be necessary are the use of disposable gloves and/or goggles.

Volatile compounds should be handled in a fume cupboard, which is certified as conforming to legal requirements (such as those laid down by the HSE in the United Kingdom). Fume cupboards should not be used with the front open above the displayed marks to ensure the correct airflow.

- Are there any aspects of the use of equipment or procedures that expose laboratory workers to any hazards? There are the ubiquitous procedures, such as pipetting. This should never be done by mouth, and must be done using either an automatic pipette or a device that can be attached to the end of a plastic or glass pipette. The instructions for operating equipment must be adhered to. Examples are centrifuges, spectrophotometers and equipment containing lasers or radiation sources.

- A vital part of a risk assessment is to identify methods of disposal of hazardous chemicals and biological materials. Many water-soluble compounds can be disposed of in a sink, usually after appropriate dilution. Lipophilic compounds and solvents are disposed of in specially designated bottles. Biological waste is usually placed in yellow bag to await later incineration. Used plastic pipettes and tips are placed in special containers, as are sharp objects such as syringe needles.

- The procedures to be taken in event of an accident or emergency must be clear. Chemical spills are a common occurrence and different procedures are required depending on the nature of the chemical. Dilute solutions of water-soluble, non-toxic chemicals are easily cleaned up by use of absorbent materials such as paper towels. All other potential hazards must be assessed, such as flammability, reactivity to air or water, corrosion or high toxicity; the incident should be immediately reported. Special measures will have to be taken. Flammable chemicals are absorbed with sawdust or special pads and the laboratory is ventilated maximally. Acids and alkalis should be diluted and neutralized.

1.2 THE LABORATORY RECORD BOOK

The importance of keeping a laboratory notebook is often underrated. Evidence collected for any purpose will not be credible if a contemporaneous record of events is not available. This is no less true for laboratory evidence than it is for police and forensic records. A book must be kept where all procedures, calculations, observations and results, along with the relevant health and safety forms, are kept. This should be a permanently bound book, and not a loose-leaf from which pages may be removed. Entries must be made contemporaneously in the laboratory at the time at which they occurred. This is frequently not appreciated by students who think that they will "write it up neatly" at some later time. This is unacceptable. For this reason, many hospital and research laboratories employ strategies such as forbidding record books to be removed from the laboratory, or insisting that duplicate records are kept and one copy left in the laboratory upon leaving at the end of the day. There are several essential pieces of information that must always be recorded.

- Entries must be done using a pen and not an erasable pencil. Corrections should be made by crossing out rather than deleted.
- Pages must be dated and the name(s) of experimenters be recorded. All entries of data on computers are date-stamped and not subject to later manipulation. Computer records should be backed up after each day to prevent loss.
- All details of methods, instruments and apparatus must be recorded. All details of chemicals and solutions (especially their concentrations) noted. Details of animals used must be available, including their species, age, weight and sex.
- Raw data must be carefully recorded and fully annotated. This includes any photographs or diagrams.
- All stages of calculations and dilutions must be written down so that any errors can later be unequivocally detected and corrected.
- Graphs and tables derived from the results should be drawn as soon as possible, preferably before leaving the laboratory. This enables an early interpretation of the results to be made, so that any adjustments in the protocol can be made before proceeding with further experimentation.

1.3 USE OF ANIMALS IN PRACTICAL PHARMACOLOGY

Even before enrolling on a pharmacology course, students must be aware that the use of living tissues and cells are integral to the discipline. Most universities post a caveat to this effect in their course descriptions. The anti-vivisectionist viewpoint is highly appreciated, and in all developed countries it is incorporated into the laws governing the use of animals in teaching and research. The use of living animals in teaching up to graduate level is not advocated. At post-graduate level and beyond, there are strict laws that must be adhered to, and licenses that must be obtained before any work on living animals can proceed. In the United Kingdom, the laws governing animal experimentation are embodied in the Animals (Scientific Procedures) Act 1986 and Amendments (2012). In addition to the Codes of Practice relating to the general care, housing and treatment of animals, both a project and personal license must be obtained from the Home Office. It is stressed that these will not be granted unless the following criteria have been justified:

- that there are no non-animal alternatives,
- that the benefits expected from the programmes of work are judged to outweigh the likely adverse effects on the animals concerned,
- that the number of animals used and their suffering must be minimized. Any contravention of these regulations found by inspectors and others will lead to a ban of an institution and individuals from working with animals.

All the experiments in this book, which is targeted at graduate-level students, do not require a licence as no substance or treatment is ever administered to animals, the numbers of animals are minimized and the above criteria laid down by the United Kingdom. Codes of Practice of the Home Office are fulfilled. Most courses do not require the use of large numbers of animals, and institutions cannot justify the expense of maintaining an animal house and attendant technicians. An alternative is to have an animal holding room where animals are delivered from breeders and held for a matter of no more than a few days. Nevertheless, animals are housed in a quiet air-conditioned room, provided with a regular light–dark cycle. Before experimentation, animals are rapidly and humanely killed (usually by cervical dislocation) in a quiet location, and tissues removed and placed in an appropriate physiological buffer to maintain their viability.

1.4 EXPERIMENTAL DESIGN

Before designing an experiment, the answers to four basic questions must be clearly understood:

- What is the topic of the experiment?
- Why is the topic being addressed?

- How is the experiment going to be carried out?
- How is the data going to be analysed?

The first two questions are answered by doing sufficient background reading, both from review articles and more detailed reports. The question of how to economically perform a literature review is discussed in the Section 10.5. Having grasped an understanding of the topic, a hypothesis can be formulated, that can be tested which will allow an advance in the understanding of the subject. Even if it is intended to repeat some previously reported preliminary finding, a full understanding of the topic is essential if a valid experiment can be designed. Importantly, there must be critical eye for detail. It is much better to attempt to design an experiment to answer one well-defined question than to attempt to address several rather vaguer questions.

How the experiment is actually designed will depend on a number of factors, including available techniques and materials. There must be a realistic estimate of both cost and the time involved. Both of these frequently underestimated, so it is wise to allow a margin of error for both of these factors. If any of the techniques are new to the experimenter, it must be certain that help will be available from a person who has first-hand knowledge of carrying out the technique. It must be borne in mind that if time is to be allowed to learn a new technique, this must be factored into the time allowed. As a general rule, in carrying out experiments and projects which have been allocated a short period of time, the learning of new methodologies should be avoided.

The actual design of an experiment should include a full understanding of the following factors.

An experiment designed to attempt to disprove the hypothesis is more powerful than one from in which it is highly likely that the results will confirm it.

The inclusion of controls is vital if unequivocal conclusions are sought. These are frequently termed "positive" and "negative" controls. A positive control is one for which a positive response is expected. They may perform the function of being "quality" controls, or may merely confirm that the technique is functioning in a predictable manner. They can also give an indication of the sensitivity of the method, and that this is sufficient for the purposes of the experiment. A negative control is always included to ensure that the method is actually measuring changes in the dependent variable (response), and will exclude any interfering variables. A negative control is sometimes referred to as a "blank", and a high variable will indicate an interfering factor.

Another factor that will determine the design of the experiment is the type of statistical analysis that will be carried out on the final results. This will then determine the number of replicates that are necessary in order that a valid statistical test can be applied (see Section 1.6).

To arrive at a final experimental design, if usually necessary to perform some pilot studies ("proof of concept"), to ensure that all is working out as planned and that there is not some flaw that has been overlooked which will confound the experiment. Typically, these are factors such as the speed of response of measuring apparatus, instability of the preparation or poor replication of results. These must be resolved before a large-scale study can begin.

1.5 UNITS, DILUTIONS AND LOGARITHMS

It may seem banal to stress the importance of understanding the basic units of mass (this is essentially the same as weight, at least on Earth), volume and concentration. Whilst all units used in pharmacology are metric, there is some variation as to whether SI units (Le Système international d'unités), or units which are derived from the fundamental units of the SI system should be used. The base units of mass, length and force defined in the SI system are the kg, m, and N, respectively. In practice, the g and mL are used in the laboratory. Similarly, The SI system prescribes that units of concentration should be expressed as kg/dm^3, whereas g/mL are commonly used in the laboratory.

Since the magnitude of each of these parameters can fall over a vast range, or orders of magnitude (an order of magnitude is generally taken to be 10-fold), they are expressed as a single digit before the decimal point multiplied by 10 to the power of a number (e.g. 0.0004 g is 4×10^{-4} g). A more convenient nomenclature is to apply a prefix to the basic unit, so that 0.0004 g can also be expressed as 0.4 mg or 400 µg. The most common prefixes for base units are graded by a factor of a thousand.

Mega (M)	10^6
Kilo (K)	10^3
milli (m)	10^{-3}
micro (µ)	10^{-6}
nano (n)	10^{-9}
pico (p)	10^{-12}
femto (f)	10^{-15}

1.5.1 Units of Mass

The common unit of the mass is gram. However, it is more useful in many cases to express weight in *moles* (or for ions, *equivalents*). The reason for this is that 1 mol of any compound contains the same number of molecules, equal to Avogadro's number, 6.022×10^{23}. In pharmacology, and biological chemistry in general, it is more useful to work in moles than grams, since it is of more interest to know the number of molecules, rather than grams, participating in a reaction or competing for a binding site, such as a receptor. Obviously, the weights of different compounds that contain 1 mol will differ hugely. For example, 1 mol of acetylcholine chloride weighs 181.66 g, and 1 mol of human acetylcholinesterase weighs 67.796 kg, yet they both contain the same number of molecules. One mole of a compound contains the molar mass in grams. The molar mass is the molecular weight (MW) in grams. The MW is sometimes called the formula weight (FW) of the relative molecular mass (RMM), and can be expressed in daltons (Da). In the case of large MW compounds such as proteins or polynucleotides, the MW is expressed in kilodaltons (KDa). Strictly, 1 Da is 1/12 of the mass of carbon-12, but in practical terms this is the same as the weight of one hydrogen atom (or proton).

The *equivalent* is also a unit of weight and is a similar concept as a mole, but expresses the number of ions in solution. An equivalent of an ion is the molar mass divided by the valency. Thus, 1 mmol Na^+ = 1 mEq Na^+, but 1 mmol Ca^{2+} = 2 mEq Ca^{2+}. An electrochemically neutral solution must contain an equal number of equivalents of positive and negative ions.

1.5.2 Units, Concentrations and Logarithms

Concentration is expressed as weight/volume, or for liquids as vol/vol. Just as weight can be expressed in grams, moles or equivalents, concentration can be expressed in a variety of units. When working in a laboratory, it is invaluable to be able to rapidly convert these different units.

Weight/volume are commonly expressed as g/L (g L^{-1}) or mol/L (mol L^{-1}) = Molar (or M). Note that 1 µg/µL = 1 mg/mL = 1 g/L and 1 µmol/µL = 1 mmol/mL = 1 mol/L = 1 Molar or 1 M.

Occasionally, concentrations are expressed as weight%, which means weight per 100 mL, and vol% means volume per 100 mL.

It is important to distinguish the commonly used molar or molarity from molality. A 1 molal solution contains 1 mol per 100 kg of solvent.

The molarity of a solution changes with temperature, since the volume of the liquid will expand, whereas molality does not.

Many drugs that are used in pharmacology experiments are expensive, indeed some pharmaceuticals are among the most expensive on Earth and can only be bought in µg quantities. Solutions of expensive compounds have to be made in small volumes. As an example, if 10 mg of tubocurarine is purchased, and a 10 mM stock solution is required, how can this be done? This is a small weight and it would be wasteful to weigh out mg quantities, and it may be preferable to calculate the volume of water that should be added to provide a 10 mM solution. First, calculate how many moles are there in the bottle.

The form in which tubocurarine is supplied is a tubocurarine hydrochloride pentahydrate with a MW of 771.72, so 10 mg = 10/771.72 mmol = 0.013 mmol or 13 µmol.

13 µmol dissolved in 13 mL gives a 1 mM solution, so 13 µmol in 1.3 mL is a 10 mM solution.

1.5.3 Dilutions

Drug dilutions are made serially, usually with a constant dilution factor. If this factor is 10, then this is a logarithmic dilution (a dilution factor of 1:10 or dilution ratio of 1:9; 1 vol of stock solution + 9 vol of diluent, for a total of 10 parts). If the dilution factor is 2 (a dilution factor of 1:2 or dilution ratio of 1:1 means 1vol. stock solution + 1vol. diluent, for a total of 2 parts). The concentration that is most important, and is cited in reports of experiments, is the final concentration present in an organ bath or enzyme reaction mixture. This is simply the volume of drug added to the organ bath or assay (V_{added}) as a fraction of the total volume (V_{total}) multiplied by the original concentration of the drug C_{stock}.

$$\text{Final concentration} = \left(\frac{V_{added}}{V_{total}} \right) \times C_{stock}$$

Particular care must be to ensure that all the units are the same. For example, if 10 µL of a 1 mg/mL solution is added to 20 mL organ bath, the same units must be used throughout. 10 µL is expressed as 0.01 mL. The final concentration in the organ bath is:

$$\left(\frac{0.01}{20} \right) \times 1 - 5 \times 10^{-4} \text{ mg/mL or 0.5 µg/mL}$$

Frequently, it is advantageous to work in molar units throughout, starting with the stock solutions. This prevents calculation errors and it is easy to arrive at the final molar concentrations.

A few examples are given to illustrate some typical calculations.

1. 30 μL of 1 mM d-tubocurarine (d-TC) was added to a 25 mL organ bath. What was the organ bath concentration?

   Answer: Final bath concentration $= (0.03/25) \times 1 = 1.2$ μM.

2. What weight of sodium chloride should be weighed out to make 20 L of 0.9% saline? What is the molarity of this solution (MW $= 58.44$)?

   Answer: 0.9% saline is 0.9 g NaCl per 100 mL of water $= 180$ g per 20 L. 0.9% means 9 g/L, so molarity of 0.9% NaCl $= 9/58.44 = 0.154$ M.

3. How would you make up 10 mL of a 50 μM solution. Assume that the smallest quantity that can be weighed on the balance is 1 mg, and make any additional solutions in a volume greater than 1 mL.

   Answer: A 50 μM solution contains 50 nmol/mL, so 10 mL contains 500 nmol $= 0.5$ μmol. MW of ACh is 181.7, so 0.5 μmol is $181.7 \times 0.5 = 90.85$ μg. This quantity is too small to be weighed on the balance, so make 10 mL of 5 mM by weighing 9 mg of ACh and dissolving this in 10 mL. Dilute 0.1 mL of 5 mM ACh in 9.9 mL water to give 10 mL of 50 μM.

4. An anti-hypertensive drug (MW 368) is recommended to be administered i.v. at 5 mg/kg. What volume of a 10 mM solution should be administered to an animal which weighs 132 g?

   Answer: Weight to be given $= 5 \times 0.132 = 0.66$ mg. To convert 10 mM to mg/mL, 10 mM of drug $= 10 \times 368 = 3680$ mg/L $= 3.68$ mg/mL. So $0.66/3.68 = 0.18$ mL should be administered to the animal.

1.5.4 Logarithms

Logarithmic relationships between variables are common in pharmacology. It is therefore important to have a clear understanding how to manipulate logarithms in calculations. When expressing numbers as logarithms, a base number must be specified. There are two commonly used base numbers, 10 ($\log_{10}$) and $e = 2.7183$ ($\log_e$, ln or natural logs). $\log_{10}$

are most widely used in pharmacology. The logarithm of a number (x) is the base number (b) to the power of that number (i.e. b^x). For example,

$$\log_b x = b^x = y$$

and

$$\log_{10} 100 = 10^2, \text{ so 2 is the } \log_{10} \text{ of } 100$$

The term antilog is more accurately described as the exponent, or the power to which the base must be raised to equal the original number.

$$\textit{anti } \log_{10} y = 10^x$$
$$\textit{anti } \log_{10} 2 = 10^2 = 100, \text{ so 100 is the antilog of 2}$$

Using a calculator, the exponent or antilog is found by using the inverse log function. The order in which numbers and functions are entered varies between different models of calculator, and should be checked for each model.

Logarithms are frequently first encountered in chemistry in the pH scale which is used to express the hydrogen ion concentration. For practical purposes, pH is defined as the negative logarithm to the base 10 of the hydrogen ion concentration $[H^+]$. Thus in a solution of pH 7, the $[H^+]$, or more accurately the hydronium ion $[H_3O^+]$ is $-$antilog $7 = 10^{-7}$ M. Note that an increase in one unit of pH scale means that the $[H^+]$ *decreases* 10-fold, because

$$\text{Change in}[H^+] = [10^{-7}] - [10^{-8}] = 10^{-1} = 0.1$$

In pharmacology, a logarithmic scale is used to express the potency of a drug, be it agonist or antagonist (see Section 2.1). If an agonist has a potency (EC_{50}) 10^{-6} M, then it is said to have a pD_2 of $-\log_{10} 10^{-6} = 6$. In order to compare the potency of this drug with another drug with a pD_2 of 8, then it may be thought that this would simply be $6/8 = 0.75$, so drug A is 0.75 times less potent than drug B. *This is incorrect!* When the log of a number is divided by the log of another, they are in fact subtracted, and the answer is the log of the difference:

$$\frac{pD_2A}{pD_2B} = (-\log 6) - (-\log 8) = \log 2 = 100$$

So drug A is in fact $100\times$ less potent than drug B.

1.6 ESSENTIAL STATISTICS

This section is not intended to give a comprehensive account of statistics as applied to biomedical experiments, since this has been done in books dedicated to the topic. For a more detailed explanation of statistical procedures, recommended books are those by Ennos (2012) and Motulsky (2003). The latter refers to the explanatory manuals (available as books and online) for GraphPad Prism, and is particularly useful since the software was written by bioscientists for bioscientists, and there are regular updates and analytical tips available on their website.

From the outset, it should be clear that there are three very different types of data:

- continuous or quantitative data
- ranked data
- non-continuous or categorical data

1.6.1 Continuous Data – t-test, ANOVA, Non-parametric Tests and Regression

Continuous (or quantitative) data consist of measurements of variables such as contraction, blood pressure temperature, time or concentration. Since the basic aim of statistics is to be able to extrapolate from sample measurements of a population to the entire population, and then assess the probability that one population (treated or untreated) differs from another. To do this it is necessary to estimate the variability of sample measurements (precision), and the distribution of individual measurements. The precision is described by the variability by descriptive statistics, the mean and standard deviation. It is vital to know whether or not the sample measurements are distributed "normally", where a bell-shaped Gaussian curve is obtained. It is simple to check the distribution of the sample measurements by simply plotting a frequency curve. It will be apparent how important it is to have a sufficient number of sample measurements. A bare minimum of 10 will start to form a pattern, but really 10 times this is needed to obtain a reliable frequency distribution curve. A table of descriptive statistics will give information about the variation of the samples in the population. This can be carried out by any statistical software package. It will consist of information about the mean and variability and confidence limits. It will convey information as to whether the distribution is symmetrical or skewed on either side of the mean. The standard deviation expresses the variation of samples around the mean, and depends how "flat" the bell-shaped curve is. If there is small variation about the mean it will reflect a sharp bell-shaped curve. The standard deviation (SD, S or σ) is defined as

$$SD = \sqrt{\frac{\sum (x - x_{mean})^2}{N - 1}}$$

where x = each value, x_{mean} = mean of values and N = number of values.

If the results are evenly distributed in a Gaussian manner, then 68% of the results fall within SD of the mean, and 95% of the values fall within 2SD of the mean.

The standard error of the mean (SEM)

$$SEM = \frac{SD}{\sqrt{N}}$$

This indicates the precision with which the true population mean has been estimated.

The appropriate statistical test to analyse quantitative data depends on the aim and design of a study. The study may have one of two basic aims.

To calculate the probability that two populations are *different* from each other, that is to say that they have different means (e.g. treated compared with control). The appropriate test will depend on (a) whether the measurements are "normally" distributed, and (b) the numbers of populations (e.g. treatments). Parametric tests are used to test normally distributed data, otherwise non-parametric tests must be used.

Comparing Two Populations

The method of testing for a difference between treatments is to propose a "null hypothesis", which states that the two populations have the same mean and distribution (i.e. that they are both from a single population). Statistical tests will calculate probability that this is true or false. If it is true, the null hypothesis is accepted, and it can be said that there is a high probability that there is no difference between the populations. It is only when the null hypothesis is rejected that it can be stated that there is probability that there is a difference. The probability level commonly taken is 95%, which can be expressed as P = 0.05 or 1 in 20. This is accepted as being the minimum level of probability that is regarded as statistically significant. For normally distributed, parametric populations, a Student's t-test can be used. To test the null hypothesis, the value of a t-statistic is calculated. This is a measure of how different the means are relative to the variability. In general, the t-statistic is calculated as the difference between the two means divided by the standard error of the difference between the means:

$$t = \frac{M_A - M_B}{SEM}$$

The t-statistic is tested against a critical value (obtained from tables or embedded in a computer program). The larger the t-statistic the greater the distance between the means, and the more likely that the null hypothesis will be rejected. The smaller the value of the t-statistic, the more likely the null hypothesis will be accepted.

For parametrically distributed populations, the t-statistic is calculated in a slightly different manner depending on the design of the experiment.

1. Comparing the average of a set of samples of a population with an expected value, such as that of a larger population.
2. Comparing two sets of *paired* values, such as before and after treatment measurements. Samples are said to be paired if they are the same animals, subjects or cells on which the control and treatment measurements are taken.
3. Comparing two sets of *unpaired* groups. The sets are unpaired if the control and treated sets are different animals, subjects or cells.

If the populations are not distributed in a Gaussian manner, a parametric test should be used. These tests compare the ranks of each sample, irrespective of the treatment group to which they belong. Examples are the Mann–Whitney test and the Wilcoxon signed-rank test.

Example of a Student's t-test

According to a report, the angiotensin receptor antagonist, losartan, inhibited platelet aggregation. This was curious because angiotensin receptors had not been reported on platelets, but if this was true it would document an additional beneficial effect of losartan, which is

Table 1.1 Inhibition of collagen-stimulated platelet aggregation (measured as decrease in optical density) by losartan.

	$\Delta E/min$
Control	Losartan
0.155	0.120
0.100	0.130
0.144	0.129
0.103	0.151
0.136	0.136
0.124	0.129
0.115	0.148
0.132	0.144
0.102	0.141
0.125	0.137

widely prescribed for hypertension and heart failure. A single concentration of 10 mM was used in the study. To investigate any inhibitory effect on a platelet agonist, a sub-maximal collagen concentration of 7.81 µg/mL was selected. A platelet aggregation experiment was carried out in which there were three control measurements, followed by three measurements in the presence of 10 mm losartan. A t-test was performed to statistically assess any differences between the treatments (to calculate the chances that the null hypothesis was true or false). The results obtained are shown in Table 1.1, and plotted as a bar graph in Figure 1.1.

An unpaired, two-tailed t-test was carried out using GraphPad Prism v.4. The options selected were unpaired since each sample was a separate sample of the platelet suspension. If each sample was tested before and after losartan, a paired test would have been selected. A two-tailed test was appropriate since there was no reason to think that the variance would extend in one direction only. The following results were obtained (see Table 1.2).

Software packages designed to perform statistical test produce a large array of data that must be interpreted. Prism is more user-friendly than many programs, such as Microsoft Excel, since it has many aids to interpretation. There are three sections to the results table – results of the unpaired t-test, a list of the differences and an F-test to test whether the variances are equal for both groups. The results of the t-test indicate that $t = 1.936$, and the degrees of freedom (d.f.) $= 20 - 2 = 18$. From this, a

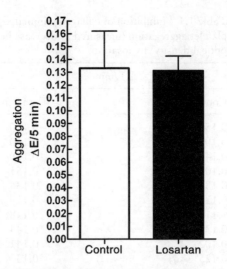

Figure 1.1 Inhibition of collagen-stimulated platelet aggregation by losartan. The large bars show the means of the two samples and the error bars indicate the SEM, $n = 10$. A t-test is carried out to test whether there is a statistical difference between the two samples.

Table 1.2 Results of the t-test of the data Table 1.1 as produced by GraphPad Prism v.4.

Parameter	Value
Column A vs. Column B	Control vs. losartan
Unpaired t-test	
P value	0.0688
P value summary	Not significant
Are means significantly different? ($P < 0.05$)	No
One- or two-tailed P value?	Two-tailed
t, d.f.	$t = 1.936$, d.f. $= 18$
How big is the difference?	
Mean $\pm$ SEM of column A	0.1236 ± 0.005916, $N = 10$
Mean $\pm$ SEM of column B	0.1365 ± 0.003067, $N = 10$
Difference between means	-0.0129 ± 0.006664
95% confidence interval	-0.02690–0.001101
R squared	0.1723
F-test to compare variances	
F, DFn, Dfd	3.722, 9, 9
P value	0.0634
P value summary	Not significant
Are variances significantly different?	No

P value of 0.0688, which is greater than 0.05 so there is more than a 95% (1 in 20) chance that the groups are the same, thus the null hypothesis is accepted. The relatively large F value suggests that the variances in the two groups are not different.

Analysis of Variance

An analysis of variance (ANOVA) compares measurements or multiple-dependent continuous variables (such as weight, enzyme activity, dose or time). For comparing two dependent variables with an independent variable, a one-way ANOVA is used. For comparing more than three variables, a two-way ANOVA is used. To illustrate the application of these tests, it may be desired to expand on the investigation as to whether a losartan inhibited platelet aggregation. In the simple experiment where a single concentration of losartan was compared with a control group without losartan, the results of a Student's t-test showed that there was no difference between the two groups. This was thought to inclusive since only one concentration of both agonist and losartan were used. If it was wished to expand the experiment by including two concentrations of losartan, three variables would have to be compared, control and the two losartan concentrations. The results of this experiment would best be analysed using a one-way ANOVA. It may be thought that a more complex test could be avoided by applying multiple t-tests, but it should be appreciated that this is not the same thing as using a one-way ANOVA. However, it was thought that since the losartan concentration selected was maximal, it may be more informative to examine the effect of one losartan concentration over the range of agonist concentrations that cover its concentration–response curve. Here the control and losartan groups are tested at seven concentrations of agonist (collagen). This now requires the use of a two-way ANOVA, followed by a Bonferroni post-test. Table 1.3 shows the results of such an experiment.

Regression

Regression analysis is used to evaluate how closely two variables are related to each other. Linear regression is used to test whether two variables are linearly correlated. Here the probability that one variable is directly proportional to the variation of another variable. Often two variables are related by a more complex relationship than a straight line. In this case, the relationship between two variables can be tested

Table 1.3   Results from an experiment where the aggregation of platelets was tested at seven different concentrations of the agonist, collagen, both in the presence and absence of losartan. Each measurement was performed in triplicate.

Collagen (μg/mL)	Aggregation (Δ E/min)					
	Control			Losartan (10 μM)		
0.00	0.077	0.065	0.062	0.047	0.050	0.050
1.95	0.610	0.057	0.074	0.067	0.080	0.083
3.90	0.092	0.117	0.160	0.083	0.099	0.036
7.81	0.155	0.100	0.144	0.120	0.130	0.143
15.60	0.366	0.404	0.416	0.222	0.241	0.288
31.25	0.458	0.495	0.499	0.283	0.378	0.409
62.50	0.480	0.500	0.463	0.274	0.369	0.354
125.0	0.464	0.580	0.472	0.225	0.341	0.298

how well the data fit a particular model. This is non-linear correlation. The most frequently encountered example of this in pharmacology is the log concentration (or dose)–response relationship which produces a sigmoid curve. A frequent misinterpretation of a correlation between two variables is that this proves causation of one event by another. In pharmacology, an increase in concentration of a drug may cause an increase in response, but an increase in response does not cause an increase in drug concentration. In this case, the response would be termed a dependent variable, and dose an independent variable; the response depends on the drug concentration, but not vice versa. Graphically, the independent variable is plotted on the horizontal axis (x-axis) and the dependent variable plotted on the vertical (y-axis). For example dose (independent) against response (dependent).

In a linear regression a straight line can be fitted to the graph by minimizing the variability of each point from a line. A straight line is described by the equation

$$y = mx + c$$

where m = the slope ($\Delta y / \Delta x$) and c is the intercept of line with y when $x = 0$.

When deciding how to plot data to which you wish to perform a linear regression, it is important to define which are the independent (x) and dependent (y) variables. A regression of x on y will not give the same result as y on x.

The more closely the two variables are related to each other is described by the Pearson coefficient (R). A coefficient of 1 indicates that the variances of the individual observations from the line are extremely small, and as this value decreases, the larger are the variances from the line. This is sometimes described as the "goodness of fit". A major use of a linear correlation graph is in predicting the magnitude of one parameter from another. A very common application is the standard curve for a spectrophotometric assay. This allows the prediction of the concentration of a substance from its absorbance by using the formula

$$y = mx + c$$

where y = absorbance of unknown sample and x = concentration of unknown sample. It is important to realize that this prediction is only valid within the absorbance range of the standards that provides a linear correlation between absorbance and concentration (see Figure 1.2).

Non-linear Correlation   For many data, the dependent variable is not linearly related to the dependent variable. Here a suitable model must be selected, and test how well the data fits this model. This may be, for example, a hyperbola (as ideally found in many binding experiments) or

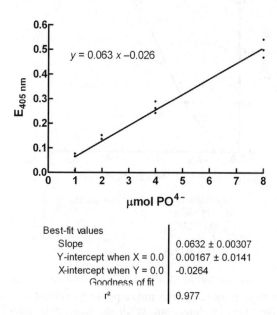

Best-fit values	
Slope	0.0632 ± 0.00307
Y-intercept when X = 0.0	0.00167 ± 0.0141
X-intercept when Y = 0.0	-0.0264
Goodness of fit	
r^2	0.977

Figure 1.2   Linear regression analysis calculates the probability that the dependent variable (y) is linearly related to the independent variable (x). The "goodness of fit" is expressed by how near the Pearson coefficient (R) is to the value of one.

an exponential growth curve. GraphPad prism actually gives the option of selecting one of a large number of models. Details of the equations for the different non-linear models, and further explanation, are given by Motulsky and Christopoulos (2003). In pharmacology, a typical model is the log concentration (or dose)–response relationship, and binding data as used in ELISA assays This is a *sigmoid curve* when the log of the independent variable (concentration) is plotted against the response (see Figure 1.3).

It is important to evaluate how well the curve fits the data. Help with doing this is given by 95% confidence interval (CI) and an R-squared value. For a good fit, narrow CI values and a high (close to 1) R^2 value is expected. Some common sense is also required in assessing the fit of the curve. Frequently, there is insufficient data to fit the curve to reasonable parameters. If the minimum value is negative or the maximum value is astronomical, this is clearly absurd, and any EC_{50} value deduced from

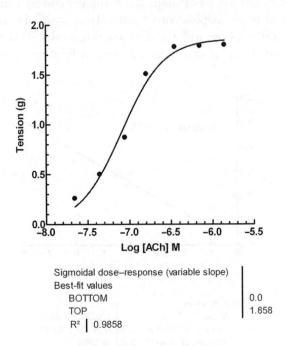

Sigmoidal dose–response (variable slope)	
Best-fit values	
BOTTOM	0.0
TOP	1.858
R²	0.9858

Figure 1.3 Non-linear correlation is useful in pharmacology for calculating the probability that the log concentration and response are related by a sigmoid curve. This provides the "best-fit" line from which the EC_{50}, maximum and minimum response values and Hill slope can be obtained. The points were fitted to the equation for a sigmoid dose–response curve, variable slope, using GraphPad Prism v.4. (GraphPad Software, Inc, San Diego, CA, USA.)

Table 1.4 Results of a Bonferroni post-test, following a two-way ANOVA.

Collagen (mg/mL)	Difference	t	P value	Summary
0.0000	−0.0190	0.6906	$P > 0.05$	Not significant
1.950	0.0130	0.4725	$P > 0.05$	Not significant
3.900	−0.05033	1.829	$P > 0.05$	Not significant
7.810	0.0006667	0.02423	$P > 0.05$	Not significant
15.60	−0.1350	4.907	$P < 0.001$	***
31.25	−0.1273	4.628	$P < 0.001$	***
62.50	−0.1487	5.404	$P < 0.001$	***
125.0	−0.1933	7.027	$P < 0.001$	***

the curve is invalid. Using GraphPad Prism, it is possible to constrain the values for maximum and minimum to fixed values, but usually it means that more data must be obtained.

Comparing concentration–response curves is problematic, as discussed by Motulsky and Christopoulos (2003). An easy approach to this is to perform a two-way ANOVA, but this can only be done with caution. A problem is that in performing ANOVA, concentration is considered as any other non-continuous treatment, like a series of drugs. When dealing with a continuous variable, like concentration or time, consideration should be given as to whether there is a trend. This can be applied to the experiment already discussed above to illustrate the use of a two-way ANOVA (Figure 1.3). Here two concentration–response curves can be drawn, one for the control curve for collagen alone, and the other for the curve in the presence of losartan. It is seen that the curve for losartan-treated platelets clearly has a lower maximum value. Applying a two-way ANOVA (Table 1.4) confirms that the points near the maximum value have a low P value ($P < 0.001$), so it seems reasonable to accept this result. This is illustrated in Figure 1.4.

1.6.2 Discontinuous Data – χ^2 and Fisher's Exact Test

These are non-continuous measurements such as the numbers of cells, tissues or animals that possess an attribute like alive or dead. In terms of digital data terminology, it is the effect of a treatment on producing one of the two states. Here contingency tables are constructed in order to carry out a Chi-squared (χ^2) test or Fisher's exact test.

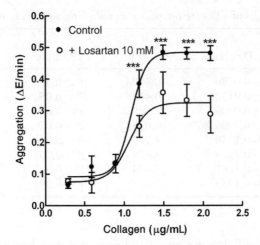

Figure 1.4 Comparison of the log concentration–response curves for control and losartan treatments, and results of a two-way ANOVA. P = <0.001 is indicated by ∗∗∗, each point shows the mean of triplicate observations, and the error bars indicate the SEM.

These tests are used to test whether a treatment significantly affected the outcome of an event in two similar groups but which were exposed to two different treatments, for example, whether treatment with a drug influenced the occurrence of a disease. They are used for analysing the difference between frequency distributions. These tests are used to analyse the so-called non-continuous data. The analysis may be by a χ^2 test or Fisher's exact comparison test. The χ^2 test is *inaccurate for small numbers* of observations, but is easier to calculate by hand. The Fisher's exact test is more accurate but involves more complex calculations.

Experiments to be analysed by χ^2 test or Fisher's exact test yield an outcome that is a categorical variable such as *alive/dead* or *disease/no disease*. To perform these tests, they must be arranged in a contingency table (see Table 1.5). This may be applied to a wide range of experimental designs.

Table 1.5 Example of a contingency table.

	Response	No response
Control		
Treatment		

There are five types of experimental designs that can be analysed using a contingency table.

- Prospective study. In this design, two closely matching groups are selected. One group is exposed to a treatment at a high dose, and the other to a lower dose. At the end of the study, the numbers in each group showing a response, such as developing cancer are noted. The results are then displayed in a contingency table before applying the appropriate statistical test.
- Cross-sectional study. Here a single, random group of subjects is selected. They are then sorted into two groups according to whether they have been subjected to a treatment. The numbers in each group showing a response is assessed, and the results displayed in a contingency table.
- Retrospective case control. This design is similar to a cross-sectional study, but may be regarded as superior since the numbers in the two groups are under more control. Two groups are selected according to whether they have been subject to a treatment, and it is then assessed whether they show a response.
- Experiment manipulating variables. This design resembles the first design, the prospective study. Two groups of closely matched subjects are selected, but in this case, one group receives a treatment and the other acts as a control. At the end of the study, the numbers in each group showing a response are assessed.
- Accuracy of a diagnostic test. Here two groups are selected according to whether they have been diagnosed with a characteristic or disease, such as AIDS for example. All subjects then undergo a diagnostic test for the disease and classified as to whether they register positive or negative in the test. The results can then be displayed in a contingency table, showing the numbers with and without the disease against whether they registered positive or negative in the test.

The χ^2 Test   Here it is determined whether the observed frequencies of an event differ significantly from the expected frequencies. It should be remembered that this test is inaccurate for small numbers in each group (say <10).

$$\chi^2 = \sum \frac{(\text{observed} - \text{expected})^2}{\text{expected}}$$

Table 1.6   Contingency table for vitamin C data.

	Survived	Died
Control	15	35
Vitamin C	31	17

The χ^2 statistic is compared with a table of χ^2 values for degrees of freedom (which is $n - 1$) against P values. Usually the value for $P = 0.05$ is taken. If the calculated χ^2 value is greater than the tabulated value, then the null hypothesis is disproved and treatment does have a protective effect at a certain probability level – if $P < 0.05$ for the χ^2 statistic at the relevant degrees of freedom, then there is a 95% probability (or chance) that the treatment produces a response.

This is best explained by an example. In a classic experiment, it was tested whether vitamin C protected guinea pigs from the effects of rabies virus infection (Banic, 1975). A random sample of guinea pigs was taken and all were infected with the same dose of rabies virus. Some were given vitamin C and others were not. The numbers of animals surviving and dying were recorded in the two groups. The results were displayed in a 2 × 2 contingency table (see Table 1.6).

Table 1.7 shows he calculation of the χ^2 value. The assumption is made that vitamin C has no effect (null hypothesis). The expected odds of dying for the two groups is calculated. Of a total of 98 animals, a total of 52 died. Therefore, the chances of dying is $(52/98) = 0.5306$ or

Table 1.7 χ^2 test of data in Table 1.6. The χ^2 value is 11.7601. This can then be compared to a table of values of χ^2 for various d.f. and P values. Note that for a 2 × 2 contingency table, d.f. $= 2 - 1 = 1$. Since obtained χ^2 value (11.76) is $> \chi^2$ for $P = 0.005$ if null hypothesis is correct (7.879), then we must reject the null hypothesis and conclude that vitamin C significantly decreases mortality rates. Usually, it is not necessary to perform these calculations since most computer software packages will perform this calculation automatically when supplied with the contingency table.

Response	Observed (0)	Expected (E)	(O – E)	(O – E)^2/E
Vit. C died	17	25.4694	8.4694	2.8163
Vit. C survived	31	22.5306	8.4694	3.1837
Crl. died	35	26.5306	8.4694	2.7037
Crl. survived	15	23.4694	8.4694	3.0564
Total	98	98		11.7601

53%. For the random sample of 48 animals in the vitamin C group, we would expect $48 \times 0.5301 = 25.469$.

Fisher's Exact Test   This is another test that can be used to statistically test the data in a contingency table, and must be used if there are small numbers in each group (<10 per cell in the table) to provide an accurate estimate of the P value. A disadvantage may be that it requires the use of a computer program since the calculation of the P value is more complex than in χ^2 test.

REFERENCES

Animals (Scientific Procedures) Act 1986 and Amendments (2012) www.legislation .gov.uk/ukdsi/2012 (accessed on 3 June 2013).

Banic, S. (1975) Prevention of rabies by vitamin C. *Nature.* 258: 153–154.

Control of Substances Harmful to Health Regulations (2002) www.hse.gov.uk/cossh (accessed on 3 June 2013).

Ennos, R. (2012) *Statistical and Data Handling Skills in Biology*, 3rd edn. Pearson Education.

Motulsky, H. (2003) *Prism 4 Statistics Guide – Statistical Analyses for Laboratory and Clinical Researchers.* San Diego, CA: GraphPad Software Inc.

Motulsky, H. and Christopoulos, A. (2003) *Fitting Models to Biological Data Using Linear and Non-linear Regression. A Practical Guide to Curve Fitting.* San Diego, CA: GraphPad Software Inc.

2

Basic Pharmacological Principles

The basic principles behind modern pharmacology as a science have their roots in the early twentieth century along with great advances in physiology and biochemistry. In addition to the terms and techniques that have been developed by biological scientists, several concepts have been developed that are peculiar to pharmacology. To emerge from the superstitious and qualitative age of medieval Materia Medica, pharmacologists needed to develop quantitative methods to determine the biological potency of compounds of both natural and synthetic origin. Only then it was possible to turn to focus on the central question of pharmacology which is to discover how drugs act on the body. The result of these endeavours was the establishment of techniques for bioassay and the advancement of theories of drug–receptor interactions.

2.1 DRUG–RECEPTOR INTERACTION

2.1.1 Agonists

From the early work of Langley (1905) and Ehrlich (1913), it became clear that some purified drugs and newly discovered neurotransmitters acted in minute quantities by combining with a "receptive substance", which then produced or inhibited a response. The first attempt to quantify the relationship between drug concentration and response was made by Hill (1909). From a study of the actions of nicotine and curare on

Practical Pharmacology for the Pharmaceutical Sciences, First Edition. D. Michael Salmon.
© 2014 John Wiley & Sons, Ltd. Published 2014 by John Wiley & Sons, Ltd.

frog rectus abdominis muscle, he found that the relationship between the agonist concentration and its response was a hyperbolic curve. This was later found to resemble the physico-chemical processes described by Langmuir (1916) for the adsorption of gases to sites on a solid surface and the relationship between the substrate concentration and the velocity of an enzyme reaction. Working on the action of acetylcholine on frog rectus abdominis muscle, Clark (1933) found that a graph of the log[ACh] against the contractile response produced a sigmoid curve. He proposed that the interaction of a drug with a receptor resembles the curves derived by Langmuir, and that it conforms to the simple model:

$$\text{drug (D)} + \text{receptor (R)} \underset{k^{-1}}{\overset{k^{+1}}{\rightleftarrows}} \text{drug} - \text{receptor (DR)} \rightarrow \text{response} \quad (2.1)$$

According to the Law of Mass Action, the value of the dissociation constant (K_d) could be derived as follows:

$$K_d = \frac{k^{-1}}{k^{+1}} = \frac{[D][R]}{DR} \quad (2.2)$$

It was fully appreciated at the time that K_d is the equilibrium binding constant, and is not a functional parameter relating drug concentration to the response. If the concentration of an agonist is plotted against the response, a hyperbolic curve is found (Figure 2.1). This is analogous to the [substrate] versus velocity relationship for an enzyme reaction. Rather one of the many transforms that are used in enzyme kinetics (e.g. a Lineweaver–Burk plot introduced in 1934), agonists were characterized by their log concentration–response relationship, which was thought to validate this model (Figure 2.1).

The potency of an agonist is defined as the EC_{50}, that is, the concentration that produces 50% of the maximum response, and the pD_2 is the negative log of the EC_{50} expressed as a molar concentration (Figure 2.1). The relative potency of two agonists A and B is EC_{50} of A/EC_{50} of B or *antilog* (pD_2 of A – pD_2 of B). The pD_2 is generally less than the pK_d, the reasons for which are explained shortly.

The maximum response elicited by a series of agonists acting at the same receptor frequently is not the same, although they all can occupy 100% of the receptor sites. A partial agonist is one that produces only a fraction of the maximum response compared to full agonist, even when 100% of the receptors are occupied. This fraction is referred to as the

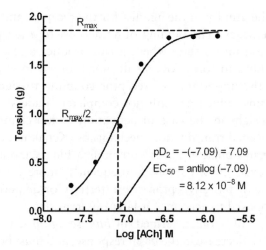

Figure 2.1 Example of a sigmoid logarithmic concentration–response curve. Note that the potency is the concentration of the agonist required to produce 50% of the maximal response. This is the antilogarithm of the value read off the x-axis, when y = maximum response/2.

"intrinsic activity" (α). Stephenson (1956) introduced the term "efficacy" to express the ratio between response and receptor occupancy. Thus a "high-efficacy" agonist can produce its maximum response even when only a fraction of the receptors are occupied, whereas a "low-efficacy" agonist cannot produce the same maximum response even when 100% of the receptors are occupied. This situation implies that the EC_{50} of an agonist is not the same as the K_d as expressed in the original model proposed by Hill and Clark. An attempt to directly determine the value of K_d is made by ligand-binding experiments, which are more fully addressed in Section 9.2.3. As eloquently described by Colquhoun (2006), the term "efficacy" remained a rather elusive concept until the report of del Castillo and Katz (1957) on the action of acetylcholine acting at the nicotinic receptor ion channel at the neuromuscular end plate of the frog rectus abdominis, a type of skeletal muscle. They proposed that the receptor could exist in two states, inactive and active. The relationship between measurements of the binding of an agonist to a receptor and its response can be more usefully expressed as:

$$
\begin{array}{ccccc}
vacant & K & occupied & & \\
A + R & \rightleftarrows & AR & \rightleftarrows & AR^* \\
inactive\,(no\;response) & & E & & active\,(response)
\end{array}
\tag{2.3}
$$

Here K is the affinity of the agonist for the receptor and E is a factor representing efficacy. The binding of an agonist to a receptor changes its conformation into a state where it can produce a response. Partial agonists can bind to form AR but do not proceed to efficiently form AR*. This is the importance of receptor coupling to signal transduction mechanisms, which are still not completely understood. There is much more insight in the case of activation of ion channels than for G-protein-mediated transduction mechanisms (Kenakin, 2006). The situation shown in Equation 2.3 is known as a binary model, where the receptor can exist in two states. Subsequently, more complex models including three or more receptor states (tertiary or quaternary models) have been proposed to account for later observations.

Another type of agonist, termed an *inverse agonist*, is also known. These produce a decrease in the basal response, and must be termed agonists, rather than antagonists since pure antagonists produce no response on their own. The action of inverse agonists is explained in terms of Equation 2.3 by postulating that in a resting state, in the absence of agonist, the receptor exists as an equilibrium between inactive R and active R*. An inverse agonist would then reduce the amount of receptor in the activated R* form. An example of an inverse agonist is the convulsive effects of some β-carbolines acing at the $GABA_A$ receptor. Agonists acting at this receptor, such as benzodiazepines, have sedative effects.

2.1.2 Antagonists

Antagonists act in one of the three ways:

1. Competitive and reversible antagonists (surmountable) compete with agonists for binding to the same receptor site in a reversible manner (e.g. atropine at muscarinic receptors)
2. Competitive but irreversible antagonists (unsurmountable) compete with the agonists for binding to a receptor-binding site, but the antagonist forms a covalent bond in a region of this site thus permanently making a proportion of the binding sites unavailable to the agonist (e.g. phenoxybenzamine at α-adrenoceptors).
3. Non-competitive antagonists are also surmountable but bind to a different site from the agonist-binding site (e.g. ketamine which binds to the ion channel of the glutamate NMDA receptor)

The first category, *competitive and reversible*, is the most amenable to kinetic analysis and is generally the most useful therapeutically. The development of the theory of competitive antagonism ran parallel with that of competitive inhibition of enzymes (Michaelis and Menton, 1913; Haldane, 1930), yet it was not until 1959 that Schild published a practical method for experimentally determining the potency of an antagonist. Gaddum (1937) and others had approached the situation of two ligands (agonist A and antagonist B) competing for the same binding site on a receptor. When B occupies binding sites on the receptor, it will reduce the sites occupied by A. This will depend on the concentrations of A and as well as the equilibrium dissociation constants of A and B. The receptor occupancy of A (ρ_A) was expressed by Gaddum as:

$$\rho_A = \frac{c_A}{1 + c_A + c_B} \tag{2.4}$$

where $c_A = \frac{[A]}{K_A}$, and $c_B = \frac{[B]}{K_B}$, and K_A and K_B are the equilibrium dissociation constants for A and B, respectively.

Schild (1947, 1949, 1957) proposed a way of measuring the affinity of an antagonist for a receptor which did not depend on knowledge of the receptor occupancy, and thus avoided the obscure relationship between binding and response. Schild only considered the concentration of the agonist that is required at various antagonist concentrations to produce the same response. This made the reasonable assumption that the same receptor occupancy by the agonist will also produce the same response. Thus, in the presence of antagonist the receptor occupancy of A (ρ_{A_1}), when the agonist concentration is [A_1]:

$$\rho_{A_1} = \frac{c_{A_1}}{1 + c_{A_1} + c_B}, \quad \text{where } c_{A_1} = \frac{[A_1]}{K_A} \tag{2.5}$$

In the absence of agonist, the receptor occupancy of A with agonist concentration [A_0] is:

$$\rho_{A_0} = \frac{c_{A_0}}{1 + c_{A_0}}, \quad \text{where } c_{A_0} = \frac{[A_0]}{K_A} \tag{2.6}$$

Since equal receptor occupancies of agonist produce equal responses, $\rho_{A_1} = \rho_{A_0}$

$$\frac{c_{A_1}}{1 + c_{A_1} + c_B} = \frac{c_{A_0}}{1 + c_0},$$

$$\text{therefore } \frac{[A_1]/K_A}{1 + [A_1]/K_A + [B]/K_B} = \frac{[A_0]/K_A}{1 + [A_0]/K_A} \quad (2.7)$$

which simplifies to $\frac{[A_1]}{[A_0]} = \frac{[B]}{K_B} + 1$, where $\frac{[A_1]}{[A_0]}$ is the ratio of concentrations of agonist in the presence of antagonist and in the absence that produce the same response. This was termed by Schild as the dose or concentration ratio (CR).

So,

$$CR - 1 = \frac{[B]}{K_B} \quad (2.8)$$

and taking logarithms,

$$\log_{10}, \ \log(CR - 1) = \log[B] - \log K_B \quad (2.9)$$

This is known as the Schild Equation. Schild was very cautious not to relate the response with the number of activated receptors, but merely assumed that equal effects involved an equal number of receptors (Arunlakshana and Schild, 1959). For this reason, he coined a term $pA_x = -\log[B]$, which was defined as the negative log of the concentration of an antagonist that will reduce potency of an antagonist x times (Schild, 1949). Schild did not state that K_B in the above was the equilibrium dissociation constant of the antagonist, but merely a constant (K) related to the production of the response. Schild plotted a graph of $\log(CR - 1)$ against pA_x. For a competitive, reversible antagonist, a straight line will be obtained with a slope of -1. When $CR = 2$, then $\log(CR - 1) = 0 = pA_x$, and was termed the pA_2 value, and is a measure of the potency of an antagonist (Figure 2.2). It can be confusing why it was necessary to use the pA_2 scale was used rather than the well-established pK_B scale. pA_2 values are based on experiments where responses are measured, whereas pK_B values pertain to measurements of receptor binding. Whilst these two values are usually very close, almost always the pK_B is slightly greater than the pA_2, and they are not interchangeable.

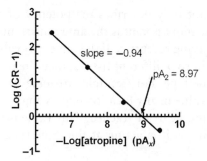

Figure 2.2 A Schild plot of –log[antagonist, in molar units] against log (CR – 1). The pA_2 value is read off the graph at the intercept of the line with x when log (CR – 1) = 0. The pA_2 value does not have any units.

If a tissue is exposed to a competitive antagonist, it will cause a parallel shift of the log concentration–response curve to an agonist (Figure 2.3). The extent to which it shifts the parallel portion of the curve is measured by the CR. This is the concentration of agonist in the presence of antagonist required to produce a fixed response on the linear part of the concentration–response curve divided by the concentration of agonist required to produce the same response in the absence of antagonist.

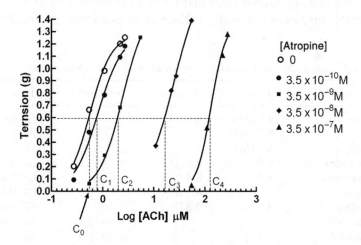

Figure 2.3 Parallel log concentration–response curves for an agonist in the absence and presence of various concentrations of antagonist. The concentration ratio (CR) is calculated from the concentrations of agonist required to produce the same response in the absence and presence of antagonist. The selected response should fall on the linear portion of the log concentration–response curve. Note that it is necessary to take the antilogarithm of the values read off the x-axis before the ratio is calculated.

The pA_2 value not only describes the potency of an antagonist (the larger the value, the more potent is the antagonist), but is a primary way of defining or classifying receptors. If the same pA_2 value is obtained for a certain antagonist in two different tissues, then the receptors are defined as being of the same type. For example, atropine would be expected to give the same pA_2 value in any location expressing muscarinic receptors, for example in ileum, heart or trachea. This value is independent of the agonist used as long as it is acting at the same type of receptor as the antagonist.

It is important to appreciate that antagonists are only relatively specific for a type of receptor, and at higher concentrations they can act at other receptor types. Thus, in many cases it is useful to know the pA_2 of an antagonist at several receptor sub-types to determine the receptor selectivity of the antagonist. This will have therapeutic implications and chances of undesirable side effects. This is evident from Table 2.1.

Note that the receptor selectivity of an antagonist is calculated by subtracting one pA_2 value from another and then taking the antilogarithm, or inverse log on a calculator (see Section 1.5.4).

Another way of estimating the K_B value of an antagonist is by using the Cheng–Prusoff relationship (Cheng and Prusoff, 1973; Craig, 1993). The EC_{50} (M) of the agonist is read from the log concentration–response curve and then a concentration that falls on the linear part of this curve (preferably greater than the EC_{50}). The responses are then recorded in the

Table 2.1 Some pA_2 values obtained for antagonists in different tissues.

Antagonist	ACh	Histamine	Isoprenaline	Tissue
Atropine	9.0 (M3)	5.6 (H_1)		GP ileum[1]
Atropine	9.2 (M2)			Frog heart[1]
Atropine	8.8 (M3)			GP bronchi[1]
Diphenhydramine	6.6	8.0 (H_1)		GP ileum[1]
Mepyramine	4.9	8.4 (H_1)		GP ileum[1]
Cimetidine		6.1 (H_2)		GP atria[1]
Cimetidine	3.8	3.2 (H_1)		GP ileum[2]
Propranolol			8.36 (β_1)	GP atria[3]
Propranolol			8.56 (β_2)	GP bronchi[3]
Atenolol			7.21 (β_1)	GP atria[3]
Atenolol			5.3 (β_2)	GP bronchi[3]

Note: The receptor sub-type is indicated by the letters within parentheses.
[1] Bowman and Rand (1980).
[2] Barker (1981).
[3] O'Donnel and Wanstall (1979).

presence of increasing concentrations of the antagonist, when a sigmoid curve is obtained. Response is plotted against log[antagonist] (M) and the IC_{50} read off this curve. The K_B is then calculated from following equation:

$$K_B = \frac{IC_{50}}{1 + (EC_{50}/[agonist])} \quad \text{(all in molar units)} \quad (2.10)$$

Unsurmountable antagonists
The determination of the affinity for a receptor of a non-competitive antagonist is more difficult, since the slope of the dose–response curve for an agonist decreases as the concentration of a non-competitive antagonist is increased. The maximum response obtained also decreases. This reflects the fact that the total number of available receptors is decreased as the antagonist concentration increases. Arunlakshana and Schild (1959) suggested that an empirical parameter, termed pA_h, could be used to describe the affinity of a competitive, irreversible antagonist. pA_h is defined as the negative logarithm of the molar concentration of a competitive, irreversible antagonist that reduces the maximum response to the agonist to one-half of the maximum response of that found in the absence of antagonist.

Furchgott (1966) suggested a novel use for competitive, irreversible antagonists to determine the affinity of an agonist (K_d). This method assumes that the receptor occupancy required to produce certain response is the same both in the absence and the presence of a competitive, irreversible antagonist. Concentration response curves for an agonist are determined alone and in the presence of the competitive antagonist. In the presence of the antagonist, the curve is not parallel and the maximum response is decreased compared with in its absence. The agonist concentration required to produce a selected response (R) in the absence of agonist is designated A, and that required to produce the same response (R) in the presence of antagonist is A*. If 1/A (ordinate) is plotted against 1/A* (abscissa), then a straight line is produced, and K_d can be calculated as:

$$K_d = \frac{(\text{slope} - 1)}{\text{intercept}} \quad (2.11)$$

The K_d determined by this method is lower than the EC_{50} value. This is accounted for the observation that the maximum response is obtained when only a fraction of the available receptors have been occupied by an agonist. This is termed the receptor reserve. It has been verified that the K_d values obtained using Furchgott analysis are very similar to those obtained by receptor-binding techniques (Morey *et al.*, 1998)

2.1.3 Receptor Classification

A crucial development of the receptor theory was the demonstration that one neurotransmitter can act on more than one type of receptor. All students of pharmacology are introduced to this idea very early in their studies. From an experimental point of view, it is helpful to be reminded of the history of this concept which is so central to pharmacology. In retrospect, it is surprising to realize that the concept of chemical transmission at neural synapses took many years to become universally recognized, at the beginning of the twentieth century. Not only did Dale and colleagues unequivocally show that acetylcholine acted as a neurotransmitter, but also it was evident that acetylcholine acted at different types of receptors at the end organs in the autonomic and somatic nervous systems to produce different responses. Two chemicals of plant origin, muscarine and nicotine, were able to mimic responses evoked by stimulation of different cholinergic nerves. Muscarine produced parasympathomimetic responses, whist nicotine activated cholinergic nerves in the somatic nervous system and produced skeletal muscle twitches. Ahlquist (1948) found that different catecholamines (adrenaline (epinephrine), noradrenaline (norepinephrine) and isoprenaline (isoproterenol)) selectively inhibited certain responses mediated by the sympathetic nervous system. Despite initial stiff resistance, he argued that two different sub-types of adrenoceptors (α and β) existed in different cell types. This was a fundamental principle which has been exploited in the development of drugs that have a relative selectivity for only a limited range of sympathetic responses, thereby eliminating a range of undesirable side effects. It transpired through the use of a range of adrenoceptor antagonists that α- and β-adrenoceptors could further be classified as α_1 and α_2, and β_1 and β_2 sub-types. This has led to the introduction of drugs such as salbutamol (a β_2-agonist) which causes bronchodilation in asthmatic subjects, but normally little effect on the heart rate and β_1-antagonists for a range of cardiovascular pathologies. Extending this principle, Sir James Black was able to discover the existence of the histamine H_2 receptor sub-type through the use of the H_2 antagonist, cimetidine, and thereby devise a revolutionary treatment for peptic ulcers. The different affinities of an antagonist for receptors and agonists found in different tissues are evident from the inspection of Table 2.1.

As the criteria that have been used to classify receptor sub-types has become more refined, their number has exploded exponentially. The bases for classification are specificity of antagonists, transduction mechanisms, protein sub-unit composition and gene sequence (Kenakin, 2006). The criteria of sub-unit composition and gene sequence have such power to identify new sub-types of receptors that many of them still lack a pharmacological function. In April 2013, 1158 genes coding for different types of receptors had been identified (IUPHAR database of receptors and ion channels, www.iuphar-db.org). The British Society of Pharmacology has also published a downloadable classification of receptors and ion channels (Alexander *et al.*, 2011). This refinement in the identification of receptors holds out great promise for the development of highly specific drugs which act on receptor sub-types that are distributed in unique locations.

2.2 BIOASSAYS

Another requirement of pharmacology is the ability to quantify the biological activity of impure mixtures containing a pharmacologically active compound. A bioassay is designed to estimate the concentration or potency of a compound by measuring it biological response. Bioassays can be used to measure the concentration of a drug, as its potency relative to another drug. Whilst many drugs can be measured by analytical chemical techniques, in many cases the exact structure of the drug is not known, or the bioactivity of the drug may not be reflected by analytical chemical techniques (e.g. peptides, isomers, small molecules). Bioassays may be performed *in vivo* (in living animals) or *in vitro*. They have an important role at all stages of drug research and development and are used to estimate the potency of a drug as well as its toxicity. An essential requirement of a bioassay is the availability of a preparation of the test substance of known standard activity. This is not necessarily a pure preparation. In the cases of many biological substances the activity of an unknown can be compared with that of an international standard, and the results expressed as international units (e.g. hormones, other mediators, clotting factors). International units of activity are measured in units required to produce a certain biological response as described in the International Pharmacopoeia. International standard preparations of a huge range of biologically active substances and pathogens are held by the National Institute of Biological Standards and Control (NIBSC). These encompass biotherapeutics, diagnostics and vaccines.

There are fundamentally two different types of bioassays – quantal (all-or-none) bioassays and graded bioassays.

- *Quantal bioassays* are where categorical data are obtained and the response is an all-or-nothing, non-variable end point, such as LD_{50} or alive or dead. These are best designed and analysed using a contingency table followed by a χ^2 test or Fisher's exact test (see 1.5.2). These are most widely used in toxicity testing.
- *Graded bioassays.* There are three basic designs of quantitative bioassays which have been devised to estimate biological activity with increasing accuracy. The pervading problem in bioassays is the change in biological response over time, and to a lesser extent the inherent reproducibility of the assay system.

2.2.1   Single-point Assays

Here a log dose–response curve for the standard is established, and the response to a single dose of unknown which falls on the linear part of this curve is determined. The concentration of a test solution of a drug can simply be calculated from the equation of a straight line:

$$\text{Response} = (\text{slope}) \times (\log \cdot \text{concentration}) + c$$

where c is the intercept of the line with the abscissa (concentration) when the value of ordinate (response) is zero. The problem with this design is that the unknowns are measured at some point later than the standards used to define the curve. Even if the unknown responses are repeated several times, this does not overcome the problem of variation of responses to the same dose of drug over time. This is a frustrating problem soon encountered by students of practical pharmacology.

2.2.2   Bracketing Assays, Three-point or 2×1 Assays

This design is an attempt to reduce the effect of biological variation with time. If a dose to a solution of unknown concentration is sandwiched between two standards, then as the tissue responsiveness varies so does that of the unknowns and standards. Responses to two doses of the standard are established on the linear part of the sigmoid log concentration–response curve. A response to the unknown is found which lies midway between that for two standard responses. These are named S1 (low concentration), S2 (high concentration) and U (unknown). These responses are then repeated preferably three or more times in a different order and a series of overlapping brackets:

S1-U-S2-U-S1-U-S2-U-S1

Here for example, four overlapping bracketed estimates would have been performed. Whilst this is an improved design over the single-point assay, it does not rule out the possibility that the volume of the drug of unknown concentration did not fall on the linear portion of its log concentration–response curve. If this were the case the assay would be flawed and yield inaccurate results. This type of bioassay is described in more detail in Section 4.2.5.

2.2.3 Multi-point Assays, Such As Four-point or 2×2 Assays

These are the most accurate design of assays, since they assure that more than one response to both unknown and standard are measured over the same time periods. A graph of log concentration against response should provide two parallel lines, demonstrating that these responses fell within the linear portion of the log concentration–response curves. Here as above, the responses to two volumes of the standard (S1 and S2) are found that fall on the linear part of the sigmoid dose–response curve. The responses to two volumes of the unknown (named U1 and U2) that also fall in the linear region of the dose–response curve are found. The ratios of the two volumes of the standard and the unknown should be the same. The responses to these volumes are then repeatedly determined in a differing order in a manner called a Latin square, an example of which is shown here:

S1	S2	U1	U2
S2	U1	U2	S1
U1	U2	S1	U1
U2	S1	U2	U1

The potency of the unknown solution relative to that of the standard can then be found by plotting the log (volume) against the response, when two parallel lines will be found. From this graph, it is possible to determine the potency of the unknown relative to that of the standard. Since the concentration of the standard is known, the concentration of the unknown can be deduced. This is explained in detail in Section 4.2.5

REFERENCES

Ahlquist, R.P. (1948) A study of the adrenotropic receptors. *Am. J. Physiol.* 153(3): 586–600.

Alexander, S.P.H., Mathie, A. and Peters, J.A. (2011) BPS guide to receptors and ion channels. *Brit. J. Pharmacol.* 164(Suppl. s1): 1–324.

Arunlakshana, O. and Schild, H.O. (1959) Some quantitative uses of drug antagonists. *Brit. J. Pharmacol.* 14: 48–58.

Barker, L.A. (1981) Histamine H_1- and muscarinic receptor antagonist activity of cimetidine and tiotidine in the guinea pig isolated ileum. *Agents and Actions.* 11: 699–705.

Bowman, W.C. and Rand, M.J. (1980) *Textbook of Pharmacology*, 2nd edn. Blackwell Scientific Publications.

Cheng, Y. and Prusoff, W.H. (1973) Relationship between the inhibition constant (K_i) and the concentration of the inhibitor which causes 50% inhibition (I50) of an enzymatic reaction. *Biochem. Pharmacol.* 22: 3099–3108.

Clark, A.J. (1933) *The Mode of Action of Drugs on Cells*, Edward Arnold.

Colquhoun, D. (2006) The quantitative analysis of drug–receptor interactions: a short history. *Trends Pharmacol. Sci.* 27(3): 149–157.

Craig, D.A. (1993) The Cheng-Prusoff relationship: something lost in translation. *Trends Pharmacol. Sci.* 14: 89–91.

del Castillo, J. and Katz, B. (1957) Interaction at end-plate receptors between different choline derivatives. *Proc. R. Soc. Lond. B. Biol. Sci.* 146: 369–381.

Ehrlich P. (1913) Chemotherapeutics: science, principles, methods and results. *Lancet* ii: 445–451.

Furchgott, R.F. (1966) The use of β-haloalkylamines in the differentiation of receptors and in the determination of dissociation constants of receptor-agonist complexes, in *Advances in Drug Research*, vol. 3 (eds N.J. Harper and A.B. Simmonds), pp. 21–55. New York: Academic Press.

Gaddum, J.H. (1937) The quantitative effects of antagonistic drugs. *J. Physiol.* 89: 7P–9P.

Hill, A.V. (1909) The mode of action of nicotine and curare determined by the form of the contraction curve and the method of temperature coefficients. *J. Physiol.* 39: 361–373.

Haldane, J.B.S. (1930) *Enzymes*. Longmans, Green and Co.

Kenakin, T. (2006) A pharmacology primer – theory, applications and methods. 2nd Ed. Elsevier.

Langley, J.N. (1905) On the reaction of cells and of nerve-endings to certain poisons, chiefly as regards the reaction of striated muscle to nicotine and to curare. *J. Physiol.* 33: 374–413.

Langmuir, I. (1916) The constitution and fundamental properties of solids and liquids. Part I. solids. *J. Am. Chem. Soc.* 38(11): 2221–2295.

Michaelis, L. and Menton, M.L. (1913) Kinetik der invertinwirkung. *Biochem. Z.* 49: 333–369.

Morey, T.E., Belardinelli, L. and Dennis, D.M. (1998) Validation of Furchgott's method to determine agonist-dependent A_1-adenosine receptor reserve in guinea-pig atrium. *Brit. J. Pharmacol.* 123: 1425–1433.

O'Donnel, S.R. and Wanstall, J.C. (1979) The importance of choice of agonist in studies designed to predict beta 2 : beta 1 adrenoceptor selectivity of antagonists from pA2 values on guinea-pig trachea and atria. *Naunyn Schmeiedebergs Arch. Pharmacol.* 308: 183–190.

Schild, H.O. (1947) pA, a new scale for the measurement of drug antagonism. *Brit. J. Pharmacol.* 2: 189–206.

Schild, H.O. (1949) pA_x and competitive antagonism. *Brit. J. Pharmacol.* 4: 277–280.

Schild, H.O. (1957) Drug antagonism and pA_x. *Pharmacol. Reveiws,* 9: 242–246.

Stephenson, R.P. (1956) A modification of receptor theory. *Brit. J. Pharmacol.* 11, 379–393.

3

Isolated Tissues and Organs

The advancement of pharmacology in the early years of the twentieth century was dependent on the development of functional, isolated organs and tissue preparations. This enabled a study of the direct effect of drugs at their site of action, which cannot be unequivocally deduced from whole body experiments. To illustrate this point, consider a drug which when administered to a patient or volunteer, results in a fall in blood pressure and an increase in heart rate. Is the primary site of action of the drug the heart or blood vessels? The cardiovascular homeostatic mechanisms that regulate blood pressure preclude a definitive site of interference by a drug. This can only be determined by examining the action of the drug on individual components of the system – namely the heart and blood vessels. Such studies led early pioneers in pharmacology to elucidate the actions of neurotransmitters in terms of drug receptors and to describe the functional anatomy of the autonomic nervous system. The establishment of the site of action of a drug at the organ and tissue level is a prerequisite before embarking on cellular, sub-cellular and molecular studies of drug action.

Although most isolated tissue preparations were introduced several years ago, they have undergone much refinement over the years to improve their accuracy and ease of use. The application of computer software to record and analyse data has been the major advance, it is now far easier for students to obtain reliable data than 20 years ago. These preparations are still widely employed to provide vital preliminary information about the pharmacological activity of candidate compounds for therapeutic applications.

Several designs of *in vitro* experiments have been developed to assess different characteristics of a drug. These are (1) to determine the type of

Practical Pharmacology for the Pharmaceutical Sciences, First Edition. D. Michael Salmon.
© 2014 John Wiley & Sons, Ltd. Published 2014 by John Wiley & Sons, Ltd.

receptor at which the drug is acting and its potency (related to affinity) at this site and (2) to establish the biological activity (potency) of a drug preparation relative to a characterized standard. These are bioassay experiments, of which there are several experimental designs (Section 2.3). The action of a drug is first determined by analysing its effect on dose–response curves. The actions of antagonists can be carried out by investigating its effect on dose–response curves of agonists, and analysing the results by using the Cheng–Prusoff relationship that was originally introduced for estimating the K_i of an enzyme inhibitor (see Section 8.2.4), or constructing Schild plots (see Section 2.1.2).

3.1 EQUIPMENT FOR *IN VITRO* EXPERIMENTS

The basic requisites for maintaining a tissue in a physiologically functional state when it is removed from an animal are that it is suspended in a suitably aerated physiological buffer at an appropriate temperature. This is done by mounting the tissue in a specialized glass organ bath. Most commonly, the tissue is bathed in the buffer, or it may be perfused through the blood vessels by the use of a suitable cannula. In order to apply several doses of the drug and observe responses, there is some means of rapidly changing the fluid in the bath. This is done by providing a large reservoir of the physiological buffer connected to the bath by gravity feed regulated by taps or clamps. Some devise for the recording of responses must be devised. In order to monitor contraction or force developed by the muscle, a suitable transducer must be connected to the tissue to convert mechanical movement into a change in voltage. A plan of the overall setup of the equipment is shown in Figure 3.1.

Two designs of transducer are used, (1) isotonic, involving changes in length of the muscle and (2) isometric, where the length of the tissue is kept constant by an opposing force. The response, in terms of the electrical signal, must be able to be recorded. In the early days this was done by kymograph that consisted of a pen that wrote on a smoked revolving cylinder. This was a tedious and messy procedure, and laborious to preserve the record. This was replaced by the use of paper recorders, which are expensive to maintain, and suffer from the advantage that the result is lost if the voltage exceeds the set voltage range, and the pen travels off the end of the paper. The use of a computer monitor to display the change in voltage, and hence muscle contraction or force development, has totally obviated deficiencies of earlier equipment. Since isolated muscle preparations contain a variety of cell types, including both muscle

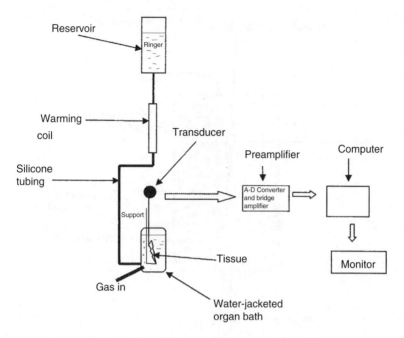

Figure 3.1 Outline of the arrangement of the equipment used for a simple organ bath experiment suitable for student laboratory use.

cells and nerves, the effect of a drug on either of these can be tested. Direct application of the drug to the tissue is usually (although not necessarily) interpreted as a direct effect on the muscle cells. If the nerve tissue is electrically stimulated by an external signal generator, then the effect of the drug on nerve function (depolarization, synthesis or release of neurotransmitter) can also be studied.

3.2 ORGAN BATHS

A variety of types and sizes of glass organ baths are available from specialized suppliers. Often the volume of the organ bath is not important, and a standard volume of 20 mL is often convenient. Small volume organ baths (1–5 mL) can be used to reduce the amount of drug administered to achieve a desired concentration. The volume of the organ bath is also important if the tissue is electrically stimulated by passing current between two electrodes placed on either side of the tissue (field stimulation), since this will affect the amount of current that is passed.

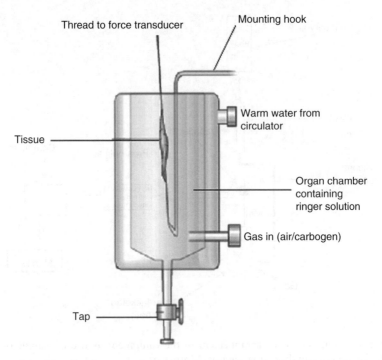

Figure 3.2   Details of an organ bath used to maintain an isolated muscle preparation.

The organ bath has a double wall to allow temperature-regulated water to warm the interior physiological buffer. The organ bath must have inlets to allow the entry and exit of water to the warming jacket, and also to allow replacement of the physiological buffer by draining and filling. The buffer is stored in a reservoir that may be as simple as a large beaker (2 L) mounted above the organ bath. A tube placed in the beaker and a siphon established to allow passage of the buffer to the organ bath below when a tap is opened. The buffer must first pass through a warming coil that may be as simple as a condenser with an outer water jacket. The outer water jacket is kept at the required temperature by circulating the water through a thermostatically regulated pump. There are various designs of organ bath, but a typical design is shown in Figure 3.2.

3.3   PHYSIOLOGICAL SALT SOLUTIONS

Many physiological salt solutions have been devised, each of which has been found empirically to maintain different isolated tissue preparations, and are frequently named according to the innovator. It is curious that for allegedly the same solution, slight variations in composition are cited in various publications (Table 3.1). The essential features are that they are all isotonic with blood plasma and are designed to maintain a pH close to 7.4. The first synthetic solution designed to maintain an isolated organ was introduced by Sydney Ringer to maintain a frog heart *in vitro*, and sometimes all similar solutions are generically described as Ringer's. Slight modifications in the composition of these salt solutions were found to be beneficial to maintain different organs. These buffers must be bubbled with different gases in order to maintain the correct pH. Thus Krebs', De Jalon's and McEwan's should be gassed with carbogen (95% O_2/5% CO_2), whilst Tyrode's is gassed with air. It will be noticed that frog's Ringer has a lower isotonicity than is used for mammalian tissues in order to mimic amphibian body fluids. It is important to note that there will be problems with the precipitation of the calcium chloride if it is not added last, and as a solution rather than the solid.

3.4 TRANSDUCERS

As mentioned above, there are two types of transducers. Isotonic transducers reflect the *changes in length* of the muscle during contraction, whilst isometric transducers record the *force* developed by the tissue when its length is maintained constant. It is notable that each of these types produces slightly different dose–response curves when used with the same tissue, with isotonic transducers giving a curve with a higher slope. Physiologically, most muscles probably display a combination of the two types of contraction, termed auxotonic. Isotonic transducers require a low DC voltage to power a small lamp focussed onto a photocell that generates a voltage. The movement of a muscle attached to one arm of a lever by a cotton thread causes the armature to move and partly occlude the beam of light impinging on the photocell. This results in a change in the voltage generated by the photocell which is proportional to the movement of the muscle. This is fed to a preamplifier and analog–digital converter for processing by the computer software. In contrast, isometric transducers record the force applied by the muscle. They operate by using four resistive strain gauges in a Wheatstone bridge arrangement. As in most pressure sensors, when a force is applied to a resistive foil there is a change in resistance. The Wheatstone bridge in an isometric transducer must be balanced at the start of the experiment such that when there is no force applied to the transducer, there is zero

Table 3.1 The composition and applications of common physiological buffers used in *in vitro* experiments. There is some variation between laboratories in the composition of some of these buffers (most noticeably in the concentrations of calcium chloride).

	Frog Ringer g/L	mM	Ringer Locke g/L	mM	Tyrode g/L	mM	Krebs–Henseleit g/L	mM	McEwan g/L	mM
NaCl	6.0–6.5	103–111	9.0	14	83.0	13.7	6.9	118	7.6	130
KCl	0.075–0.14	1–1.9	0.4	5.4	0.2	2.7	0.35	4.7	0.42	5.6
CaCl$_2$	0.1–0.12	0.9–1.1	0.12–0.24	1.1–2.2	0.24	2.2	0.28	2.5	0.24	2.2
MgCl$_2$					0.01–0.05	0.1–0.5				
MgSO$_4 \cdot 7H_2O$							0.29	1.2		
NaH$_2$PO$_4$	0.01	0.1			0.05	0.4	0.16	1.2	0.14	1.0
NaHCO$_3$	0.1–0.2	1.2–2.4	0.15–0.5	1.8–6.0	1.0	11.9	2.1	25	2.1	25
Glucose	0–2.0	0–11.1	1.0–2.0	5.5–11.1	1.0	5.5	2.0	11.1	2.0	11.1
Aeration	Air		Air or O$_2$		Air or O$_2$ or 5%CO$_2$ + 95%O$_2$		5%CO$_2$ + 5%O$_2$		5%CO$_2$ + 95%O$_2$	
Uses	Frog and other amphibian tissues		Mammalian perfused heart		Mammalian isolated smooth muscle		Mammalian tissues requiring good oxygenation (nerve, heart)			

Source: Adapted from Bowman and Rand (1980). Bowman, W.C. and Rand, M.J. (1980) *Textbook of Pharmacology*, 2nd edn, Section 4.18. Blackwell Scientific Publications.

voltage across the Wheatstone bridge. As force is applied, the bridge becomes unbalanced and a voltage develops across the bridge that can then be amplified and displayed to record tension. The isometric force transducer should be capable of recording 0.1–20 g (1–200 mN). Low cost commercial transducers are widely available, and examples are the Grass FT3 and ADInstruments MLT050 and MLT001 transducers. The transducer should be supplied with a suitable bridge amplifier.

The signal displayed on the monitor from the transducer should be calibrated to convert mV to appropriate units. In the case of isotonic transducers the response can be merely expressed as millimetre deflection from zero. Calibration of isometric transducers is a more critical procedure that must be carried out each day. Firstly, the voltage across the Wheatstone bridge must be set to zero millivolt by manipulating a zero knob or following the detailed instructions for the software. Then the voltage across the bridge is calibrated in terms of grams (or Newtons) by suspending a known weight from the arm of the transducer and supplying the conversion factor for millivolt to gram, so that the units on the ordinate on the display is labelled in the correct units.

3.5 RECORDING EQUIPMENT AND SOFTWARE

The unpopularity of *in vitro* pharmacology experiments amongst students has now been greatly ameliorated by the use of computers running user-friendly software. Almost all laboratories now use a system where the transducer is connected via an analog–digital converter and preamplifier to a computer to allow processing by appropriate software. The system most widely used is the PowerLab® range supplied by ADInstruments, which is to be recommended for its simplicity and flexibility of operation. The most basic of the Powerlab range most widely used in teaching laboratories consists of a two channel system, but four and eight channel amplifiers are also available for more advanced applications. PowerLab® amplifiers with integrated bridge amplifiers are also available. The Chart® software supplied by ADInstruments operates in a Microsoft Windows or Apple OS environment, allowing students to learn to use the system rapidly and enjoyably. Students' results can be removed from the laboratory on a flash memory device that can be analysed later using Chart Reader® software. The detailed operating instructions are supplied with the equipment. An important feature of the system is that it functions by sampling the signal voltage from the transducer with time. The minimum rate is 10–20 samples for each response. Too low a sampling rate will give angular, inaccurate record-ings, and too high a sample rate gives large files that use more computer memory. The software easily allows the responses on the monitor to be displayed by altering the amplitude (full scale, or screen deflection) and time (equivalent of chart speed) and the signal is never frustratingly lost, as with earlier recording devices. Examples of the output from typ-ical experiments recorded by Chart® software are shown in subsequent chapters in this book. A manual description of the use of ADInstruments PowerLab equipment for pharmacology experiments is available online and can be downloaded (Bowers *et al.*, 1999).

3.6 DOSING

Test substances must be administered in a consistent manner if repro-ducible and accurate measurements are to be obtained. Tissues may be maintained *in vitro* in one of the three basic arrangements. They may be mounted in a conventional (or static) organ bath of fixed volume, and the physiological buffer can be periodically replaced. This has a major advantage that the tissue is exposed to a known drug concentration (although the exact concentration to which the receptor is exposed is uncertain). Alternatively, the tissue can be superfused with a constantly flowing stream of buffer, or constantly perfused through blood vessels. Superfusion and perfusion often maintain the tissues in viable state for longer periods, but suffer from the drawback that the exact concentra-tion of drug to which the tissue is exposed is not known, and the amount of drug added is expressed only in terms of a dose (a mass or weight, in grams or moles).

Using the conventional (static) organ bath, there are two ways of constructing a concentration–response curve to drugs, (1) a drug is added to the tissue and allowed a fixed contact time with the tissue, followed by washout, and (2) cumulative drug addition. In this case, the drug is exposed to the tissue for a fixed contact time, after which there is no washout but a further drug addition (usually twice the concentration) is made. This is usually continued until no further increase in response is noted when the bath drug concentration is increased, indicating that the maximum response has been reached. A limitation of the use of cumulative dose–response curves is when sequential additions of the same dose of a drug results in a decreasing response. This is explained by desensitization or tachyphylaxis (see Section 4.1.2).

The volume of additions relative to that of the organ bath must be kept low, preferably <2% of the bath volume, that is, <0.4 mL of addition if a 20 mL organ bath is used. A number of factors should be noted.

1. The concentration required to obtain a maximum response should be determined in a preliminary study.
2. Since the concentration is logarithmically related to the response, it is best to increase the concentration by arithmetically doubling the concentration, for example, 1, 2, 4, 8 nM...
3. Additions should be added rapidly, taking care to avoid contact of concentrated solutions directly with the tissue.
4. A consistent time cycle of contact time to drug washout, followed by further drug addition must be adhered to. For example, a 2 min time cycle may consist of adding the drug at zero time and allow 30 s contact time, wash out and wait 90 s for recovery to baseline, then make second addition. The time cycle used depends on the drug administered and the type of tissue preparation used. In Chapter 4, the time cycle is specified for each type of tissue preparation. When performing cumulative concentration–response curves, the time cycle is simpler since there is no washout, but a constant time between additions must still be maintained.
5. Where the drug is washed out, an equal number of washes (usually three) must be used.
6. In superfused and perfused tissue preparations, there is an automatic washout of drug, and the time of addition of subsequent drug additions is determined by the time taken for the response to decay to baseline levels.

In static organ bath preparations, the concentration of drug in the organ bath $[D_{OB}]$ is calculated as the amount (weight) of drug added divided by the volume of the organ bath (V_{OB}). The amount of drug added is the concentration of the stock drug solution $[D_{stock}]$ multiplied by the volume added (V_{added}). Algebraically this can be expressed as:

$$[D_{OB}] = \frac{[D_{stock}] \cdot V_{added}}{V_{OB}} \qquad (3.1)$$

(Obviously, when constructing a concentration–response curve, this calculation need only be done for the first addition, as the bath concentration of all subsequent additions is merely twice the preceding one).

3.7 ELECTRICALLY STIMULATED PREPARATIONS

Nerve stimulation of isolated tissues predates the birth of pharmacology as a discipline. The Italian scientists, Luigi Galvani and AlessandroVolta, were amongst the first to study the effects of electricity on the twitching of muscles in the eighteenth century. At the beginning of the twentieth century, Einthoven introduced the use of a galvanometer to record the electrical activity of the heart, and hence the development of the electro-cardiogram or ECG. Otto Loewi discovered that the neurotransmitter acetylcholine was released from the vagus nerve to slow the rate of the frog's heart.

Experiments using electrodes to stimulate nerves in isolated prepa-rations are more technically demanding than simply adding exogenous chemicals to preparations and observing responses, but as many drugs act to interfere with the release of neurotransmitters from nerve endings, nerve-stimulated preparations are essential components in the teaching of *in vitro* pharmacology. There are several methods of nerve stimu-lation, including direct nerve and transmural and field stimulation. In order to set up preparation in which nerves are stimulated directly, it is necessary to have access to nerves so that an electrode may be attached. A well-known example of this is the Finkleman preparation of the rabbit intestine, where sympathetic nerves supplying the intestine run along the mesenteric arteries supplying the intestine. However in the intestine, postganglionic parasympathetic nerves are not accessible as they are located within the tissue in the Messner's and Auerbach's plexuses. Here methods were devised to pass a uniform electrical field across the tissue. In the transmural stimulation method introduced by Paton, one electrode was placed in the lumen of a segment of the intes-tine and the other in the physiological fluid bathing the tissue. An alter-native method more commonly employed is field stimulation, which has been applied to both isolated ileum and tracheal (Carlyle, 1964) preparations.

Electrical stimulation of isolated preparations requires the use of a stimulator. These are designed to deliver fully controlled square waves. These may be free standing units or in the case of ADInstruments stim-ulators can be integrated with the PowerLab® equipment and Chart® software. Three basic parameters that are controlled by a stimulator are

1. Pulse rate or frequency (Hz)
2. Pulse width or duration of pulse (s)
3. Stimulus intensity or amplitude (V)

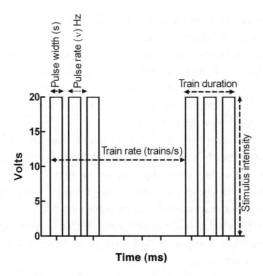

Figure 3.3   Illustration of the parameters that must be adjusted on a simple square-wave generator used to stimulate nerves or muscles for use in pharmacology experiments.

More advanced, but lesser used, functions are:

4. Delay, or time between the initiation of a pulse from an external source and the generation of the pulse (s)
5. To deliver trains of pulses (e.g. 5 pulses every 10 s) three parameters need to be set: frequency of pulses within the train, train duration and train delay (time between each train of pulses). These parameters are illustrated in Figure 3.3.

3.8   FAULT-FINDING OF *IN VITRO* ISOLATED TISSUE PREPARATIONS

1. No response on monitor
   (a) Check settings, is chart advance (start) switched on?
   (b) Does transducer produce a signal? Gently touch transducer or add 1 g weight.
   (c) Is transducer calibration correct?
   (d) Is cursor on scale? Is baseline displayed on monitor?
2. No response to drugs
   (a) Does baseline look suspiciously flat? Check cotton threads tightly fixed to support.

(b) Incorrect resting tension – usually too much.

(c) Check bath temperature.

(d) Ringer solution incorrectly made up? Is it completely clear? Replace if cloudy due calcium precipitation.

(e) Faulty electrode connections? Check contacts.

3. Very small responses which are not dose dependent

(a) Too much resting tension.

(b) Incorrect calibration of transducer? Check by adding 1 g weight to transducer. Isometric transducer not balanced.

(c) Incorrect sensitivity range set on monitor.

(d) Faulty dilution of agonist? Make up new dilution.

4. Irreproducible responses to same dose of agonist

(a) Poor pipetting technique? Check that pipetting is accurate. Check that drug is added to top of Ringer solution, but not directly on tissue.

(b) Agonist is showing tachyphylaxis. Try a different agonist.

5. Graded responses to low doses, but give square-topped responses of same size at higher doses

(a) Isotonic transducer incorrectly set up? Lever hits stop at larger responses.

(b) Mechanical obstruction to the movement of tissue in organ bath.

6. Rising or falling baseline and responses erratic

(a) Incorrect resting tension.

(b) Unstable transducer (isotonic transducers are temperature dependent).

(c) Bad electrical contacts. Check cables and plugs.

7. Irregular or noisy baseline or "spiking"

(a) Spontaneous activity of tissue. Check bath temperature.

(b) Due to particular tissue preparation – set up another tissue.

REFERENCES

Bowers, M., Briggs, D. and Purves, R. (1999) Pharmacology experiments manual (ML006). From www.adinstruments.com (accessed 6 September 2013). ADInstruments Pty Ltd., NSW 2154, Australia.

Bowman, W.C. and Rand, M.J. (1980) *Textbook of Pharmacology*, 2nd edn, Section 4.18. Blackwell Scientific Publications.

Carlyle, R.F. (1964) The responses of the guinea-pig isolated intact trachea to transmural stimulation and the release of an acetylcholine-like substance under conditions of rest and stimulation. *Brit. J. Pharmacol. Chemotherap.* 22: 126–36.

4

Smooth Muscle Preparations

4.1 GASTROINTESTINAL SMOOTH MUSCLE PREPARATIONS

The classic guinea pig ileum (GPI) preparation is typical of intestinal smooth muscle and is the most useful and economical preparation for use in undergraduate student laboratory classes. Responses of intestinal smooth muscle may be elicited either by adding an agonist directly to the organ bath, or by simulating action potentials by electrical stimulation of the tissue resulting in neurotransmitter release. The upper part of the intestine is the short duodenum, followed by the jejunum and the lower part, the ileum. Intestinal smooth muscles are innervated by the enteric, sympathetic and parasympathetic divisions of the autonomic nervous system. The main excitatory transmitter is acetylcholine that is released from the parasympathetic and enteric fibres, and acts on muscarinic (M_3) receptors on the smooth muscle cells. These muscarinic fibres originate from two ganglia, the myenteric (Auerbach's) plexus and the submucosal (Meissner's) plexus. The former is located between the longitudinal and circular muscle layers. Most frequently it is the contraction of the longitudinal fibres that is monitored. The major inhibitory transmitter is noradrenaline released from the sympathetic nerves that are located periarterially in the mesentery. A characteristic feature of the intestinal smooth muscle is the spontaneous, myogenic response, the magnitude and regularity of which depend on the region of the intestine, the species and the temperature. To study excitatory responses it is desirable to minimize this spontaneous activity. This is usually achieved by using the caecal end of the ileum obtained from a guinea pig and maintaining

Practical Pharmacology for the Pharmaceutical Sciences, First Edition. D. Michael Salmon.
© 2014 John Wiley & Sons, Ltd. Published 2014 by John Wiley & Sons, Ltd.

the tissue below body temperature at 32°C. Inhibitory activity (relaxation) of the agonists is studied in preparations that show spontaneous or electrically stimulated preparations. The two major groups of receptor sub-types that show such inhibitory activity are adrenoceptors and opioid receptors. The preparations that have been most widely used are the isolated jejunum from the rabbit that show regular and uniform spontaneous pendular activity, and the electrically stimulated guinea pig ileum.

4.2 GUINEA PIG ISOLATED ILEUM

The ileum strip from the guinea pig is the hardy workhorse known to generations of pharmacology students, and probably is the best known *in vitro* isolated tissue preparation. It has gained this reputation for several reasons. It shows relatively rapid responses and is relatively cheap, since more than 30 preparations can be set up using an ileum from just one guinea pig. It is an excellent preparation to demonstrate and quantify drugs acting on a wide variety of receptor sub-types. This is because the ileum is bestowed with a complex innervation, in addition to responses to a wide variety of autacoids and hormones. To appreciate this fully, it is useful to refer a pharmacology or physiology textbook. The innervation is typical in those tissues supplied by the enteric nervous system, such as the stomach, duodenum, small intestine and caecum. Briefly, the sympathetic innervations are relatively simple, since the post-ganglionic nerve arises from the paravertebral and terminates on the smooth muscle cells of the circular and longitudinal smooth muscles. However, the parasympathetic system is more complex. The preganglionic, nicotinic nerve that emerges from the spinal column terminates in one of the two enteric plexuses, the myenteric (Auerbach's plexus) of the sub-mucosal (Meissner's) plexus. The post-ganglionic nerve is cholinergic and terminates at the junction with muscarinic receptors.

Protocol

Experimental Conditions

Ringer solution	Tyrodes
Temperature	32°C
Organ bath volume	20 mL
Aeration	Air
Transducer type	Isometric or isotonic
Resting tension	1 g isometric or isotonic

Equilibration time	15–30 min
Dose cycle	90 s
Contact time	30 s

This classic preparation is based on that first described by Magnus (1904). Guinea pigs should be allowed to accommodate in the premises and fasted for 24 h before use. The animal is killed by cervical dislocation. An incision in the abdomen is made to expose the full length of the intestine. To locate the ileum, find the caecum that is notably thicker than the rest of the small intestine. Follow this up the small intestine to the colon where it abruptly becomes thinner. The ileum begins 1–2 cm further back. Make a cut with scissors at this point and follow this for about 40 cm where another cut is made. The ileum must be treated with care, without applying pressure or tearing as this will damage the tissue. Gently pulling the tissue will remove it from most of the surrounding mesentery. It is important to proceed rapidly at this point as any delay will result in tissue death. The ileum is placed in warm Tyrode solution gently gassed with air. Discard 2 cm next to the cut ends and place cotton ligature at one end to allow sequential sections to be removed for mounting in organ bath. The tissue will remain viable in this state for several hours. A 1–2 cm length of ileum is placed in a petri dish containing Tyrode's. If any digested food remains in the lumen, it may be removed by gently pipetting with Tyrode's. Lengths of cotton (about 40 cm) are threaded through the lumen and out through the intestinal wall at the top and bottom of the ileum. Attach one to the support in the organ bath and the other to the transducer. Ensure that the Tyrodes in the organ bath is gently bubbled with air. After a short period of warming to 32°C and equilibration, the preparation is ready to start the experiment.

4.2.1   Concentration–Response Curves for Cholinesters

Compounds producing a response when administered to a tissue are termed agonists. The magnitude of the response produced by an agonist acting directly on receptors will vary according to the dose administered. All drugs have a threshold dose below which no response can be detected. As the dose of drug administered is increased above the threshold the response evoked is also increased until it reaches a maximum value. Above this concentration, regardless of how much the dose is then increased, the response will not increase further. This is the maximum response. The potency of an agonist (EC_{50}) is found by determining the concentration that produces half the maximum response. It is more useful to express potency as a pD$_2$ value, given that pD$_2$ = $-$log EC$_{50}$ (as a molar concentration).

Differences in concentration–response curves are readily demonstrated by comparing curves for a series of structurally related cholinesters: acetylcholine, acetyl-β-methylcholine (methacholine) and butyrylcholine.

Protocol

1. From 50 μM stock solutions of each of these cholinesters, a concentration–response curve can be constructed by adding 10 μL to a 20 mL organ bath and following a strict time cycle, such as 30 s contact time, followed by three washouts with Tyrode's solution.
2. After 60 s this procedure is repeated, but at each addition doubling the volume of cholinester added until the response (tension) is not seen to increase after doubling the volume of stock solution.
3. This is done for all cholinesters in turn.
4. A typical record of the results is shown in Figure 4.1.

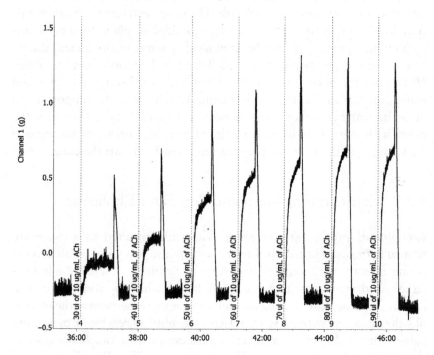

Figure 4.1 A record of responses to increasing concentrations of acetylcholine. The abscissa is time (min). Responses were recorded using Chart® software and displayed on a computer monitor. (ADInstruments Ltd., U.K.)

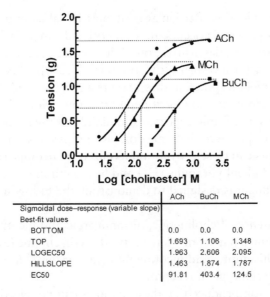

	ACh	BuCh	MCh
Sigmoidal dose–response (variable slope)			
Best-fit values			
BOTTOM	0.0	0.0	0.0
TOP	1.693	1.106	1.348
LOGEC50	1.963	2.606	2.095
HILLSLOPE	1.463	1.874	1.787
EC50	91.81	403.4	124.5

Figure 4.2 Log concentration–response curves for cholinesters.

The results in Figure 4.1 can be displayed graphically by plotting the log concentration versus response (Figure 4.2).

Questions

1. The responses are plotted against the log[cholinester] (M). A series of sigmoid curves with parallel slopes should be obtained.
2. Calculate the ED_{50} and pD_2 values. Also tabulate the maximal responses for each curve.
3. Place the cholinesters in order of relative potency and efficacy.
4. Which of the cholinesters can be termed a full agonist?

4.2.2 Selective Antagonism

Antagonists that selectively bind to receptor types have been pharmacologically useful in many ways, and have been instrumental in defining types of receptors. As the specificity of antagonists binding to different populations of receptors has been identified, the number of sub-types of receptors has grown. The guinea pig ileum contains both cholinergic receptors (M_3) and histamine (H_1) on the smooth muscle membrane. Nicotinic receptors are present in the neuronal ganglia. The agonists that selectively bind to the receptors used in this experiment are methacholine, histamine and nicotine. Particular care must be taken in using nicotine, to which tachyphylaxis occurs after repetitive dosing. Tachyphylaxis is a form of short-term desensitization of the nicotinic receptor, and can be thought of as a form of protective response. Tachyphylaxis is common with ion channel receptors, and represents a reversion of the channel to an inactive form after prolonged depolarization, with delayed recovery to the resting state. Rapid repeated stimulation of the receptor results in a decrease in response to the same dose. Responses to stimulation of different receptors can be selectively antagonized by atropine (non-selective muscarinic), diphenhydramine (H_1) and hexamethonium (neuronal N).

This experiment is designed to demonstrate the following:

1. The presence of cholinergic, histaminergic and nicotinic receptors in the guinea pig ileum. Here acetyl-β-methacholine (methacholine, MCh), histamine and nicotine bitartrate are used as selective agonists of each type of receptor.
2. Contraction produced by these agonists can be selectively antagonized by specific antagonists.
3. Hexamethonium can inhibit responses to nicotine and methacholine, but not histamine. Atropine and diphenhydramine selectively inhibit M (non-selective) and H_1 receptors, but neither antagonize responses to nicotine. Thus deductions can be drawn about the anatomical locations of these receptors.

Protocol

1. The guinea pig ileum is set up as detailed above.
2. A partial (four-point) log concentration–response curve is obtained for each agonist. For MCh, a stock solution of 5×10^{-5} M is required. For nicotine, 10^{-4} M is used. Construct a concentration–response curve for each agonist by adding 0.1 mL, and doubling the volume on subsequent additions until a maximum response is reached. *Note that the MCh and nicotine doses must be alternated in order to avoid tachyphylaxis of the responses to nicotine.* A contact time of 30 s and cycle time of 90 s should be used throughout these experiments.
3. Select a concentration that is approximately in the middle of each curve (sub-max concentrations). Obtain responses to each of these sub-max concentrations. Add hexamethonium to a bath concentration of 10^{-6} M.

4. Leave for exactly 1 min and then add the sub-maximal concentration of MCh. Wash out, and add hexamethonium again. After 1 min, add nicotine.

5. Repeat sub-maximal doses of MCh and Nicotine until original responses are restored. This ensures that hexamethonium has been washed out.

6. Now add 0.1 mL of atropine (10^{-6} M). After 1 min, add the sub-maximal concentration of MCh. Without washing out, add 1 mL of 10 μg/mL MCh.

7. Wash out and replace the atropine.

8. Wash out, replace atropine, wait 1 min, then add nicotine.

Typical results for this experiment are shown in Figure 4.3.

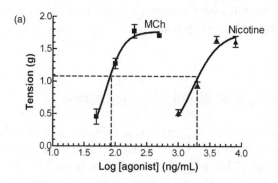

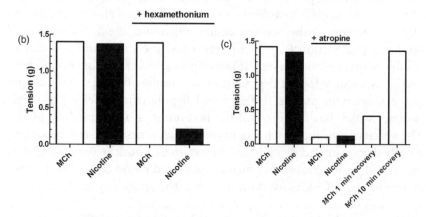

Figure 4.3 Typical results obtained for the selective antagonism experiment. (a) Concentration–response curves for methacholine and nicotine. (b) Hexamethonium (10^{-6} M) inhibited the response to nicotine but did not affect those to methacholine and (c) Atropine (10^{-6} M) inhibited responses to both methacholine and nicotine.

Questions

1. Plot the log concentration–response curves for MCh and nicotine on the same graph (use molar concentrations). What is the reason for using molar concentrations rather than μg/mL? Note the relative potencies.
2. Plot bar graphs of the responses to each of the agonists in the absence and presence of hexamethonium, diphenhydramine and atropine (label with correct bath concentrations).
3. What can you conclude about the specificities of each of the antagonists, and the receptors present in the preparation?
4. Draw a diagram of the suggested location of each of these receptors in the guinea pig ileum.
5. Define the following terms: tachyphylaxis, selective antagonism and competitive antagonism.

4.2.3 Specificity of Blood Cholinesterases

Cholinesterases exist as two isoenzymes: acetylcholine specific cholinesterase (AChE) and the broad specificity butyryl or pseudo-cholinesterase (BuChE or pChE). AChE is bound to the outer post-synaptic membrane and is also found in extraordinarily high concentration within a small portion of the mass of the electroplaques of the organ of the electric eel (*Electrophorus electricus*). There is also an exceptionally high activity at the neuromuscular junction in the dorsal muscle of the leech (see Section 6.2). pChE has a low specificity and is found in soluble form in blood plasma. It hydrolyses a wide range of cholinesters, and is responsible for the degradation of a number of drugs. An example is the short-term muscle relaxant and depolarizing blocker, succinyl-choline, which has a short half-life because it is hydrolysed by pChE. The specificity of these two isoenzymes of cholinesterase can be conveniently demonstrated using blood as a source of both enzymes. AChE is associated with extracellular membrane of red blood cells (RBCs), and, as mentioned, pChE is found in plasma. The specificity of these isoenzymes can be demonstrated by incubating the cholinesters acetylcholine (ACh), methacholine (MCh) and carbachol (CCh) with RBC and plasma and the hydrolysis of the esters monitored by observing the responses of guinea pig isolated ileum. AChE hydrolyses not only ACh, but also pChE. The carbamoyl ester, carbachol, is resistant to hydrolysis by both enzymes.

Table 4.1 Preparation of test-tubes for an experiment to demonstrate the specificity of cholinesterases for cholinesters.

Tube Number	1 mL cholinester	1 mL RBC	1 mL plasma	0.9% saline	0.1 mL eserine
1	ACh	−	−	1 mL	−
2	ACh	+	−	−	−
3	ACh	−	+	−	−
4	MCh	−	−	1 mL	−
5	MCh	+	−	−	−
6	MCh	−	+	−	−
7	CCh	−	−	1 mL	−
8	CCh	+	−	−	−
9	CCh	−	+	−	−
10	−	+	−	1 mL	−
11	−	−	+	1 mL	−
12	−	−	−	2 mL	+
13	−	+	−	1 mL	+
14	−	−	+	1 mL	+
15	ACh	+	−	−	+
16	ACh	−	+	−	+

Protocol

1. Dilute RBC suspension and plasma 1:10 with 0.9% saline. Prepare 20 µM eserine (physostigmine).
2. The isolated GPI is set up in an organ bath aerated with Tyrode's solution. Concentration–response curves for ACh, MCh, CCh are established as done previously (see Section 4.1.1). A sub-maximal concentration (administered in 0.4 mL or less) is selected from these log concentration–response curves that gives a response on the linear portion of the curve (between 30% and 70% of the maximum response).
3. Label 16 plastic test tubes and pipette the following into each tube (Table 4.1).
4. Incubate the tubes in a water bath at 37°C for 30 min. Remove from water bath and place on ice.
5. Test the response of the ileum to each of these incubated tubes using twice the volume that was used for the selected sub-maximal concentration of cholinester.

Results

6. Plot the log concentration–response curves for each of the cholinesters. Indicate the concentration selected for the incubations.

7. Plot bar graphs of the responses to the contents of each of the tubes.

Questions

1. What was the purpose of including tubes 10–14?
2. Describe and explain your results, including the action of eserine.
3. Do RBC, plasma or eserine alone produce a response in the ileum? If so, how do you account for this?
4. What responses would you predict for butyrylcholine (BuCh) alone, BuCh + RBC and BuCh + plasma?

4.2.4 Quantification of the Potency of an Antagonist

This has been introduced in Section 2.1.2. Arunlakshana and Schild (1959) defined a term to describe the potency of an antagonist in terms of the extent to which the antagonist can reduce the potency of an agonist to produce a response. Schild only considered the concentration of the agonist that is required at various antagonist concentrations to produce the same response. He introduced a term pA_x, which they defined as the negative log of the concentration of an antagonist that will reduce potency of an antagonist x times. If a tissue is exposed to a competitive antagonist, it will cause a parallel shift of the concentration–response curve to an agonist. The extent to which it shifts the curve is measured by the concentration ratio (CR). This is the concentration of agonist in the presence of antagonist required to produce a fixed response on the linear part of the concentration–response curve divided by the concentration of agonist required to produce the same response in the absence of antagonist. When the antagonist occupies 50% of the receptors, theoretically CR = 2. Schild developed a graphical method to determine a value for $-\log K_B$ that he termed the pA_2 value. The $-\log$ value is used to provide a convenient scale for the potency of an antagonist that generally ranges between 5 and 10 (equivalent to concentrations of 10^{-5}–10^{-10} M). Note that the pA_2 value has no units. An experiment is carried out to construct concentration–response curves for an agonist in the absence and presence of at least three concentrations of antagonist. From an equation derived by Gaddum (1937) for the receptor occupancy of an agonist in the presence of a competing, Schild derived an equation relating the CR to K_B (the Schild equation):

$$CR = 1 + \frac{[\text{antagonist}]}{K_B} \tag{4.1}$$

where K_B is the equilibrium dissociation constant for the binding of antagonist for the receptor. It is the concentration of antagonist that would occupy 50% of the receptors in the absence of agonist. Theoretically, $pK_B = pA_2$

And when expressed as $\log_{10}$,

$$-\log K_B = pK_B = pA_2 \tag{4.2}$$

so

$$\log(CR - 1) = -\log[\text{antagonist}] + pK_B, \tag{4.3}$$

(which is in the form of $y = mx + c$).

A graph of $\log[\text{antagonist}]$, (x), against $\log(CR - 1)$, (y), should provide a straight line.

For a competitive antagonist, the slope of the line will be 1, and if the negative $\log[\text{antagonist}]$ is plotted against $\log(CR - 1)$ gives a slope of -1. When $CR = 2$, the antagonist theoretically occupies 50% of the receptor sites, and so $\log(1) = 0$, so the intercept of the line on the x-axis $(-\log[\text{antagonist}]) = pA_2$. Note that this analysis is only valid when the slope of the Schild plot $= -1$. The pA_2 value describes the potency of an antagonist (the larger the value, the lower the antagonist concentration, and the more potent is the antagonist). This was an early way of defining or classifying receptors. If an antagonist acts on a receptor with the same pA_2 value in different tissues, then the receptors in the two sites are defined as being of the same type. For example, atropine would be expected to give the same pA_2 value in any location expressing muscarinic receptors, for example, in ileum, heart or trachea. This value is independent of the agonist used as long as it is acting at the same type of receptor as the antagonist.

Protocol

1. Prepare stock solutions of acetylcholine with concentrations of 1, 10, and 100 μg/mL.
2. Using a 90 s cycle time, construct a concentration–response curve of acetylcholine as in Section 4.2.1.
3. Now add the antagonist, atropine, to the reservoir to produce a bath concentration of 0.1 μg/mL. Empty and fill the organ bath three times to ensure that the atropine solution is in contact with the tissue. It is useful to ensure that this is the case by introducing

a small air bubble into the delivery tube, but be careful to maintain the siphon.

4. After 15 min, repeat the concentration–response curve for acetylcholine.

5. Increase the concentration of atropine in the reservoir to 1 µg/mL, and ensure that this solution is in the organ bath, as before. Leave for a further 15 min.

6. Repeat the concentration–response curve for acetylcholine.

7. Repeat this procedure once or twice more, each time increasing the atropine concentration by 10 times.

Analysis

1. Draw the log concentration–response curves for acetylcholine on the same graph.

2. Select a response that falls on the linear part of all the concentration–response curves. Read off the concentrations required to produce this response at each of the concentrations of atropine. These are denoted as C_0, C_1, C_2, etc.

3. The concentration shift ratio (CR) is calculated as C_1/C_0, C_2/C_0 etc.

4. Tabulate these values as shown in Table 4.2.

5. Construct a Schild plot by drawing a graph of $\log(CR - 1)$ on the abscissa (x-axis) against $-\log[\text{atropine}]$ on the ordinate (y-axis) as shown in Figure 2.3.

Questions

1. What is the pA_2 value for atropine in this preparation? How is this defined?

Table 4.2   Used to calculate parameters for a Schild plot.

[Atropine] µg/mL	[Atropine] M	−Log[atropine] M	CR	(CR − 1)	Log(CR − 1)

2. Is there any evidence to indicate whether atropine is acting competitively at a receptor site in this preparation? If so, which receptor site?
3. Does this agree with values cited in the literature (see Table 2.1)?

4.2.5 Bioassays

A bioassay is designed to estimate the concentration or potency of a compound by measuring its biological response (Rang *et al.*, 2012). Bioassays can be used to measure the concentration of a drug, its potency relative to another drug or its binding constant. Whilst many drugs can be measured by analytical chemical techniques, in many cases the exact structure of the drug is not known, or the activity of the drug may not be reflected by analytical chemical techniques (e.g. peptides, isomers, small molecules). Bioassays may be performed *in vivo* (in living animals) or *in vitro*. An essential requirement of a bioassay is the availability of a preparation of the test substance of known standard activity. This is not necessarily a pure preparation. In the cases of many biological substances the activity of an unknown can be compared with that of an international standard, and the results expressed as international units (e.g. hormones or other mediators, clotting factors). There are two different types of bioassays – quantal (all-or-none) bioassays and graded bioassays.

Quantal bioassays are where categorical data are obtained and the response is a non-variable end point, such as LD_{50} or alive or dead. These are best designed and analysed using a contingency table followed by a χ^2 test or Fisher's exact test (see Section 1.6.2).

Graded bioassays. There are three basic designs of graded, quantitative bioassays which can be used to estimate activity with increasing accuracy.

1. *Single-point assays.* Here a log dose–response curve for the standard is established, and the response to a single dose of unknown is compared with this curve.
2. *Bracketing assays, three-point or 2×1 assays.* Responses to two doses of the standard are established on the linear part of the sigmoid log dose–response curve. A response to the unknown is found, which is midway between the two standard responses. These are named 'std1' (low concentration), 'std2' (high concentration) and 'unk' (unknown). These responses are brackets. Here, for example, 'four bracketing' estimates would have been performed:

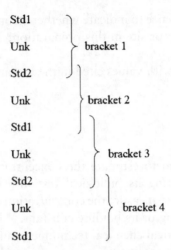

A typical set of responses are shown in Figure 4.4.

3. *Multi-point assays, such as the 4 × 4 assay.* These are the most accurate designs of assays. Here as above, two responses to the standard are found that fall on the linear part of the sigmoid dose–response curve (S1 and S2). Two responses to the unknown that

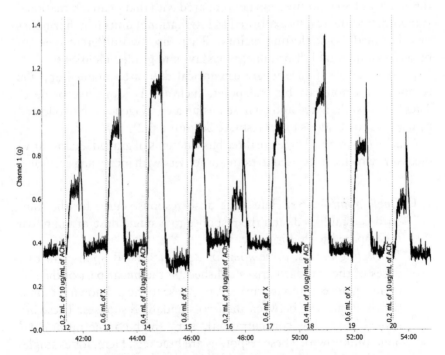

Figure 4.4   A record of responses obtained in a bracketing or three-point assay. The data were recorded using Chart software. (ADInstruments Ltd., U.K.)

also fall in the linear region of the dose–response curve are found (named U1 and U2). These responses are then repeated in a differing order.

Protocol

The guinea pig isolated ileum preparation is set up, as done for the guinea pig isolated ileum preparations described above. Check the maximum response to the standard (add 0.8 mL of 10 µg/mL ACh in the organ bath).

Single-point Assay

1. You are supplied with a standard solution and a stock concentration of the unknown.
2. Add increasing concentration of standard (use 1 µg/mL and 10 µg/mL) to the tissue until you reach the maximal response and you can plot a full log concentration–response curve.
3. Note the linear part of the curve and the approximate 50% of maximum response. Now add volumes of unknown to the organ bath aiming to find a volume (V_{unk}) that elicits a response on the linear part of the concentration–response curve. You may have to dilute the stock solution of unknown in order to obtain a response that falls on the concentration–response curve for the standard.
4. Read the exact response (R_{unk}) from the curve and find the organ bath concentration of the unknown (C_{unk}) on the abscissa that corresponds to this response.

So V_{unk} contains ($C_{unk} \times 20$) ng of standard (if organ bath volume = 20 mL).

So the concentration of diluted solution of unknown contains: $\frac{C_{std}}{V_{unk}} \times 20$

$$\frac{C_{unk} \times 20}{V_{unk}} = \text{µg/mL of unknown} \tag{4.4}$$

So if the original stock solution of unknown was diluted x times, then the original solution contains (diluted unknown concentration/x) µg/mL.

The design of this assay suffers a great deal of biological variation, and so is inherently inaccurate. This is because the standards and unknown

samples were tested over different periods of time. Isolated tissue preparations are notorious for the changes in response over time. The response
tends to gradually increase as the tissue recovers from the trauma and
changes in temperature occur during the preparation of the tissue. A
superior design is the 'bracketing assay' as described in the next section.

Bracketing (or three-point) Assay

Select two concentrations of the standard that fall in the linear part
of the log concentration–response curve, between 30% and 70% of
maximum response to the standard (std1 and std2). Now find a volume
of unknown (V_{unk}) that produces a response approximately midway
between these two standard responses (unk). The response to V_{unk} is
bracketed between the responses to std1 and std2. Now repeat these
responses in the following order to obtain four overlapping brackets:

<u>std1</u>-unk-<u>std2</u>-unk-<u>std1</u>-unk-<u>std2</u>-unk-std1

A typical record of the results is as shown in Figure 4.4.

Plot the log concentration against response for each of the brackets,
and read off the organ bath concentration obtained when V_{unk} was added
(Figure 4.5).

The calculations are very similar to those for the single-point assay.
If the unknown stock solution was diluted, the concentration of the
original stock concentration of the unknown is calculated as for the

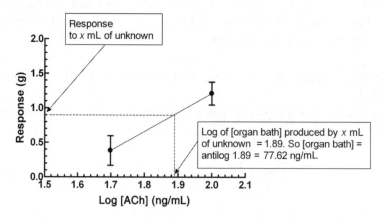

Figure 4.5 Log concentration against response for std1 and std2. The response for
the unknown is read off the abscissa to obtain the log of the bath concentration
obtained on addition of V_{unk}.

single-point assay. This will provide four estimates of the concentration of the original unknown solution expressed in µg/mL. Find the mean ± standard deviation of these values. The concentration of the original stock solution of the unknown are quoted as the mean with the coefficient of variation (CV).

$$\%CV = \left(\frac{SD}{mean}\right) \times 100 \qquad (4.5)$$

Four-point Bioassay

This is a biological assay that attempts to determine the concentration of an unknown solution by comparison of the unknown with a standard solution. The assay is designed to minimize inaccuracies introduced due to variations in response of the system occurring during the assay (biological variation). The assay is called a *2 × 2 assay* because 2 doses of the standard and 2 doses of the unknown are used. Since the standard and unknown solutions contain the same drug, the dose–response curves will be identical, and the two experimental curves (for the standard and unknown) will be parallel.

Let us say that the drug to be assayed is acetylcholine, and we have a standard solution of 100 µg/mL. A guinea pig ileum preparation is set up as normal. The maximal response of the preparation to ACh is determined (usually obtained using in excess of 0.2 mL of a 100 µg/mL solution. Ensure that the response to this dose is on scale (the sensitivity of the equipment must be adjusted if necessary). Two sub-maximal responses to ACh are found which lie on the linear portion of the dose–response curve (i.e. between 20% and 80% of maximal), using a dilution of the standard solution. These responses are designated A and B. A should be about 30% of maximum and B about 60% of maximum. Two similar responses to a dilution of the unknown are also obtained, and designated C and D. *It is important that the dose volume ratio of A:B = dose volume ratio of C:D.*

If A = 0.1 mL of a 10^{-2} (or 1:100) dilution of the standard,

and B = 0.2 mL of a 10^{-2} dilution of standard,

the dose volume ratio of A:B = 0.1:0.2 = 1:2

then in this case, the dose volume ratio of C:D must be 1:2 also (e.g. 0.1:0.2; 0.15:0.3 etc.).

The four doses ABC and D are then repetitively tested in different order in the form of a Latin Square.

ACBD

CBDA

BDAC

DACB

Calculations

To illustrate the calculations for this assay, an example of typical results is shown.

Let us say that

A = 0.1 mL of a 10^{-2} (or S/100) dilution of standard
B = 0.2 mL of a 10^{-2} (or S/100) dilution of standard
C = 0.2 mL of a 10^{-1} (or X/10) dilution of unknown
D = 0.4 mL of a 10^{-1} (or X/10) dilution of unknown.

The means of the four responses obtained for each dose are found to be as follows.

A = 40 mm; B = 83 mm; C = 33 mm and D = 84 mm

A graph of log (dose volume) against response is constructed. The two lines joining A to B and C to D should be parallel, and it is apparent that the standard is more potent than the unknown. To draw the graph it may be simpler to convert mL to μL to avoid complications when taking logarithms. Tabulate results in Table 4.3.

Table 4.3   Record of results for a 2 × 2 bioassay.

Sample	Dilution	Volume of diluted stock	μL of original stock	Log (μL original stock)	Response (mm)
A (std)	1:100	0.1 mL (= 100 μL)	100/100 = 1	0.0	40
B (std)	1:100	0.2 mL (= 200 μL)	2	0.301	83
C (unknown)	1:10	0.2 mL	20	1.301	33
D (unknown)	1:10	0.4 mL	40	1.602	84

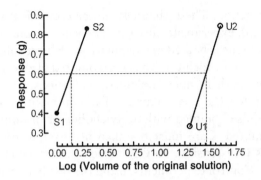

Figure 4.6 Log of the volume (in μL) of the original solution for S1, S2, U1 and U2 plotted against response.

A graph of log volume against response is plotted as shown in Figure 4.6.

From the graph, the relative potency of the unknown relative to the standard can be calculated. The relative potency is the ratio of the dose volumes of standard and unknown required to produce an equal response. For example, to obtain a response of 60 mm requires:

antilog (0.14) μL = 1.38 μL of standard and antilog (1.48) μL = 30.2 μL of unknown.

Therefore, 1.38 μL standard produces the same response as 30.2 μL of unknown.

Therefore, the unknown is $(1.38/30.2) = 0.0456$ times less concentrated than the standard, but the original standard is known to contain 100 μg/mL.

Therefore, the original unknown solution contains (0.0456×100) μg/mL = 4.56 μg/mL.

Thus, the unknown solution contains 4.56 μg/mL.

As for the bracketing assay, four estimates for the concentration of the unknown are found. The final estimate for the concentration of the unknown solution can be expressed as the mean ± %CV.

4.2.6 Calcium Channel Blockers

Maintained contraction of intestinal smooth muscle depends on depolarization of the smooth muscle cells which results in the opening of voltage-operated L-type calcium channels. The influx of calcium ions results in contraction. These channels can be conveniently studied in a simple organ bath experiment since Ca^{2+}-dependent contractions can be obtained if the tissue is bathed in a depolarizing Ringer solution containing a high $[K^+]$ from which Ca^{2+} has been omitted (Table 4.4). As $CaCl_2$ is added to the bath in micromolar concentrations, a concentration–response curve for Ca^{2+} is obtained. The contractions are slower to develop than those obtained with acetylcholine or histamine, so a little patience is required. Care must be taken not to add too much $CaCl_2$ as the tissue is very sensitive to Ca^{2+} when the L-type channels are in the open state. Control experiments, using the fast Na^+ channel blocker tetrodotoxin, have shown that contraction is not due to depolarization of nerves thereby causing neurotransmitter release. The affinity of antagonists acting at the L-type calcium channels, such as nifedipine, verapamil or diltiazem can be estimated in such a preparation. According to Spedding (1982), it can be demonstrated that nifedipine acts as a Ca^{2+} channel antagonist in an apparently competitive manner, and an 'apparent' pA_2 value determined. However, this type of antagonism of ion channel opening differs from competitive antagonism at a receptor site. This technique can be used to screen for channel blocking activity of any compound, although additional experiments must be done to define its mechanism. Control experiments should be performed as some compounds can antagonize the action of calcium-induced contractions at sites other than at the voltage-operated channel. For example, trifluoperazine acts as an antagonist in this experiment because it binds to intracellular calmodulin. If the antagonism cannot be reversed within 30 min this suggests that there is an intracellular site of action (Spedding, 1982).

Protocol

Experimental Conditions

Organ bath volume	20 mL
Ringer solution	high $[K^+]$, zero $[Ca^{2+}]$ Tyrode's
Aeration	Air
Temperature	32°C
Transducer	Isotonic or isometric
Dosing	Cumulative
Contact time	90 s or 2 min
Chart speed	0.5 cm/min

$CaCl_2$ stock solutions: 1 mM, 10 mM, 100 mM and 1 M. 1 M is prepared by dissolving 14.9 g $CaCl_2$ in 100 mL distilled H_2O. Dilutions

Table 4.4 Composition of the high [K$^+$], zero [Ca^{2+}] Tyrode's solution compared with normal Tyrode's.

	Normal Tyrode's [mM]	High-K$^+$ Tyrode's [mM]
NaCl	137	97
KCl	5	40
NaHCO$_3$	11.9	11.9
NaH$_2$PO$_4$	0.4	0.4
CaCl$_2$	2.5	0.0
MgCl$_2$	1.0	0.0
D-glucose	5.5	5.5

of 1:10 are prepared in distilled water. Nifedipine stock solutions of 10^{-8} M, 10^{-7} M and 10^{-6} M. These are prepared by dilution of a stock solution of 10^{-4} M nifedipine (3.46 mg dissolved in 1 mL DMSO and made up to 100 mL with distilled water). Solutions should be stored in dark bottles as nifedipine is light sensitive.

Procedure – The apparent pA$_2$ for the antagonism of Ca^{2+}-dependent contractions by nifedipine can be estimated by the method of Arunlakshana and Schild (1959). Obviously the depolarized tissue preparation is very sensitive to extracellular Ca^{2+} and other divalent cations. It is therefore imperative to use clean apparatus, so it is important to wash out apparatus with distilled water before you start.

1. Construct a cumulative dose–response curve for CaCl$_2$ by adding the lowest concentration of CaCl$_2$. Allow a contact time of 3 min. Without washing out, add the next dose of CaCl$_2$. Obviously, to calculate the bath concentration the total amounts of CaCl$_2$ must be added together. The ED$_{50}$ normally is approximately 0.5 mM, and approaches a maximum at less than 2 mM CaCl$_2$, but do not attempt to establish the maximum as high [Ca^{2+}] are toxic to the tissue.
2. Add nifedipine to 1 L 'high K$^+$' Tyrode's solution to produce a bath concentration of 10^{-10} M nifedipine and leave for 15 min. Repeat the cumulative dose–response curve to CaCl$_2$. If necessary, increase the bath concentration of Ca^{2+} until contractions of approximately equal size as those obtained in the previous curve are obtained.
3. Increase the concentration of Nifedipine to the beaker to a concentration of 5×10^{-10} M. Wait for 15 min and repeat the above procedure.

4. Repeat this with doses of nifedipine of 10^{-9} M, 5×10^{-9} M and if time 10^{-8} M.

Analysis

Plot the concentration–response curves (Figure 4.7a). The results can be analysed by the method described in Section 4.2.4. Calculate the dose–shift ratios (CR) by reading off the bath concentration of Ca^{2+} required to produce a fixed response that falls on the parallel part of the curves. Plot $\log(CR - 1)$ against $-\log[\text{nifedipine, (M)}]$. A straight line with a slope of -1 ± 0.2 should be produced if nifedipine acts as a competitive antagonist (Figure 4.7b). As this experiment is time consuming, it may not be possible to obtain a full set of curves; it is still possible to estimate the K_B for nifedipine by substituting values for CR and [antagonist] in the Schild equation. The 'apparent' pA_2 and pA_{10} values (when $CR = 2$ and 10 respectively) can also be read off the graph by considering that when $\log(CR - 1) = 0$, $-\log[A] = pA_2$, and when $\log(CR - 1) = 0.95$, $-\log[A] = pA_{10}$. An apparently competitive antagonism will produce a value for $pA_2 - pA_{10} = 0.95 \pm 0.2$, and the slope should be close to -1.

Questions

1. Why is the term 'apparent' pA_2 used?
2. Comment on the significance of these results in the light of what is known about the mode of action of nifedipine.
3. What experiments could help to distinguish a calcium antagonist acting at the L-type Ca^{2+} from one acting at an intracellular site?
4. What control experiment could be done to ensure that the high $[K^+]$ buffer was not acting by depolarizing nerves and releasing neurotransmitter?

4.2.7 Field-stimulated Guinea Pig Isolated Ileum

Protocol

Conditions for Electrical Stimulation

Voltage	20 V
Frequency	0.1–200 Hz
Pulse width	0.5 ms

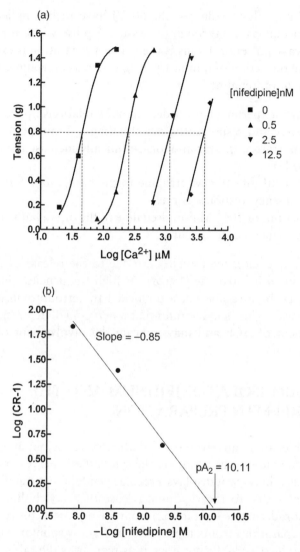

Figure 4.7 Schild plot of the antagonism of calcium-dependent contractions by nifedipine in the guinea-pig isolated ileum.

A section of guinea pig ileum is set up as in the previous experiments. Electrodes are then set up so that a current may be passed through the Tyrode's solution and across the tissue. These are usually made out of platinum, and can be of various designs. One well suited to this purpose consists of two parallel plates. As the resistance of the salt solution is far less than that of the tissue, most of the current passes directly through the salt solution, high voltages (20–60 V) have to be applied so that sufficient current passes through the tissue. A pulse width (or duration of square-wave pulse) of 0.5 ms is used. This preparation is excellent to demonstrate the action of inhibitory drugs, such as morphine, and other opioids and noradrenaline.

1. The first experiment is to determine a relationship between the frequency and response (twitches). Using a 3 min period of stimulation, start at 0.1 Hz and logarithmically increase the frequency to 200 Hz.
2. Add 0.1 mL of 10^{-5} M atropine to the bath. After 5 min, repeat the frequency–response curve.
3. Add 0.1 mL of 10^{-5} M noradrenaline to the organ bath, and repeat the frequency–response curve.

At low frequencies, the twitches are due to the release of ACh, and will be inhibited by atropine (1 µM). At high frequencies, the twitches are unaffected by atropine but are blocked by tetrodotoxin (too toxic for student use). This demonstrates that a nerve-mediated response that is independent of ACh and may indicate the involvement of another neurotransmitter.

4.3   RABBIT ISOLATED JEJUNUM AND THE FINKLEMAN PREPARATION

Whilst rabbits are relatively expensive, the isolated jejunum is an excellent tissue to study drugs acting on the sympathetic nerves and adrenoceptors. This is because it displays a regular pendular contractile activity. This is largely due to the rhythmic release of acetylcholine (contraction) and noradrenaline (relaxation) from the myenteric nerves. Isolated intestine preparations from other species also show spontaneous activity when maintained at 37°C, but none is as regular as the rabbit jejunum. A slightly more demanding preparation is Finkleman preparation (Finkleman, 1930), where the intestine is electrically stimulated periarterially. In this preparation, a short length of jejunum, with the mesentery and mesenteric artery attached, is mounted on a Saxby electrode. This consists of a plastic rod containing a metal wire that terminates in a wire ring to which the mesentery is attached. The release of noradrenaline from sympathetic nerves is frequency dependent (1–50 Hz). The preparation is particularly useful for demonstrating the action of indirectly acting sympathomimetics (tyramine, ephedrine) and sympatholytics adrenergic neurone blockers such as guanethidine).

4.3.1 Adrenoceptor Sub-types

The experiment described here is designed to illustrate the specificity of agonists and antagonists for the α- and β-adrenoceptor sub-types as established by Ahlquist (1948). The agonists that are used are adrenaline (epinephrine), noradrenaline (norepinephrine) and isoprenaline (isoproterenol). The order of specificity for the α-adrenoceptor is noradrenaline>adrenaline>>isoprenaline, and for the β-adrenoceptor is isoprenaline>noradrenaline>>adrenaline. The antagonists are phentolamine (α) and propranolol (β).

Protocol

Experimental Conditions

Organ bath volume	20 mL
Aeration	95% O_2/5% CO_2
Ringer solution	Krebs–Henseleit
Bath temperature	37°C
Transducer	Isotonic
Resting tension	Variable, <4 g
Equilibration period	30 min
Contact time	30 s
Dose cycle	4–5 min
Electrode	A Saxby electrode is used for the Finkleman preparation.
Stimulus parameters	Pulse width 1 ms
	Voltage Supramaximal
	Frequency 1–50 Hz
Stimulus duration	30 s

The preparation is set up in a very similar manner to the guinea pig ileum. Shortly after preparation is placed in the organ bath, and it has warmed up to 37°C, it is seen that the tissue shows spontaneous rhythmic cycles of contraction and relaxation. When the pendular motion becomes regular, it is possible to proceed with the experiment.

1. Cumulative concentration–response curves for adrenaline, noradrenaline and isoprenaline are established. The response to these adrenoceptor agonists is seen as a diminution in the pendular motion of the tissue. After each agonist, the organ bath is washed

out three times and a period of 5 min is required to re-establish the pendular motion again.

2. Phentolamine is then pipetted into the Ringer reservoir above the organ bath to achieve a concentration of 10^{-7} M. Fill and empty the organ bath with the Ringer–antagonist solution several times to ensure that the tissue is exposed to the antagonist. Each antagonist should be equilibrated for 10 min before proceeding to repeat the cumulative concentration–response curves as previously.

3. After the three concentration–response curves, the procedure is repeated with propranolol at 10^{-6} M. After an equilibration period of 10 min, the three concentration–response curves are repeated for the final time.

Questions

1. Plot the log concentration–response curves for adrenaline, noradrenaline and isoprenaline (see Figure 4.8). Each graph should have three curves, one for agonist alone, and two others for agonist in the presence of phentolamine or propranolol. Fit the best line for the sigmoid dose–response (variable slope) relationship.

2. Tabulate the EC_{50} values for each of the curves, and calculate the concentration–shift ratio.

3. Comment on the relative potencies of the three agonists for adrenoceptors in the rabbit jejunum. What deductions can be made about the distribution of the adrenoceptor sub-types present in this tissue?

4.4   ISOLATED TRACHEAL RINGS

Protocol

Experimental Conditions

Organ bath volume	20 mL
Ringer solution	Krebs
Aeration	95% O_2/5% CO_2
Bath temperature	37°C
Transducer	Isometric
Resting tension	0.5 g
Equilibration time	60 min
Dosing cycle time	20 min
Contact time	2 min

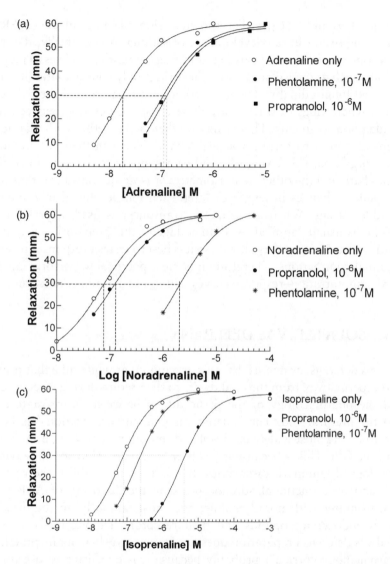

Figure 4.8 Effect of phentolamine and propranolol on the log concentration–response curves for adrenaline, noradrenaline and isoprenaline in the rabbit isolated jejunum.

Isolated tracheal preparations from the guinea pig are useful for testing compounds acting on bronchoconstriction and dilation (potentially useful to treat asthma). The trachea is innervated by the parasympathetic nervous system (contractile) and sympathetic nervous system (relaxing) and notably has a high expression of β_2-receptors. There are also contractile responses to 5-HT, histamine and the anaphylactic leukotrienes,

LTC_4, LTD_4 and LTE_4, earlier known as slow-reacting substance (SRS) that is important in the development of asthma. Anatomically, the trachea consists of a series of cartilaginous rings. As with blood vessel preparations, such as the isolated aorta, these rings only contract laterally and are fixed longitudinally. The most common preparation is the tracheal chain, where rings of trachea are dissected out by cutting between the cartilaginous segments. These rings are then tied together with platinum loops or cotton to form a tracheal 'chain'. This can then be mounted in an organ bath by fixing one end to a support at the bottom of the organ bath and the other to an isometric or isotonic transducer that can respond to changes in length or lateral tension developed in response to added drugs. When an isometric transducer is used, a preload of 200 mg is usually applied. A disadvantage of this preparation is that it contracts and relaxes very slowly, so it is best to restrict responses to submaximal contractions. Therefore, it is most practical to perform cumulative concentration–response curves and to use a 5 min contact time.

4.5   ISOLATED VAS DEFERENS

The vas deferens, or ductus deferens, is a muscular white tube that transports sperm away from the epididymis to the urethra during ejaculation, and the pharmacology of the tissue differs between the prostate end and the urethral end. The isolated vas deferens preparation has been said to be the most widely used isolated smooth preparation, even more than the GPI. This is because it has been instrumental in so many pharmacological landmark discoveries. It was used to establish the role of pre- and post-junctional adrenoceptors, in the discovery of ATP as a co-transmitter with noradrenaline, and most famously, in the discovery of endogenous opioids and their receptors. Despite this, the isolated vas deferens preparation does not feature largely in many practical pharmacology curricula probably because it is expensive to use (only two per animal) and it is more difficult to set up that the GPI. The predominant innervation is sympathetic, which is atypical, because the ganglia are located close to the tissue and there are only short, post-ganglionic nerves. α-Adrenoceptors appear to be predominantly present, with post-synaptic, excitatory α_1-receptors and pre-synaptic inhibitory α_2-receptors, although β_2-receptors are also present. Indirectly acting sympathomimetics, such as tyramine and cocaine, cause contraction. There is evidence for a sparse parasympathetic innervation, but of more importance are non-adrenergic, non-cholinergic (NANC) transmitters and modulators, namely purines, peptides such as vasoactive polypeptide and nitric oxide. The distribution of opioid receptors depends on the species. The mouse vas deferens was used by Hughes and Kosterliz in their pioneering studies in the early 1970s because the μ- and δ-receptors in this preparations are much more sensitive to enkephalins than the guinea pig ileum (Waterfield et $al.$, 1977).

Protocol for the Mouse Isolated, Field-stimulated Vas Deferens

Experimental Conditions

Organ bath	5 mL or less
Ringer solution	Mg^{2+}-free Krebs
Aeration	95% CO_2/5% O_2
Temperature	37°C
Transducer	Isometric
Resting tension	0.2 g
Equilibration	10 min
Dose cycle	10 min
Electrical stimulation	
Voltage	>100 V
Pulse width	1 ms
Frequency	0.1 Hz
Electrodes	Platinum plates on either side of the tissue

In order to study inhibitory activity in this preparation, it is normally electrically stimulated. This can be done either by using the vas deferens–hypogastric nerve preparation, or by field stimulation. The latter is described here.

The vas deferens is easily identifiable as muscular, thin, fine tubes in close association with an artery. One end is cut close to the epididymis, and the other end is cut close to the urethra. This is placed in Krebs Ringer (without Mg^{2+}) in a petri dish, and all the adhering mesentery is removed. A thread is tied round both ends, one of which is attached to a metal support and the other left free to be tied to the transducer.

QUESTIONS ON ISOLATED TISSUE PREPARATIONS

1. An experiment was performed to determine the dose–response of the guinea pig isolated ileum to three muscarinic receptor agonists, A, B and C. An organ bath containing 15 mL of Tyrode's solution was used. The following data shown in Table 4.5 were obtained.

Table 4.5 Data obtained for concentration response data for choline esters in guinea-pig isolated ileum.

Addition to organ bath	[Agonist] μg/mL	Response to A (g)	Response to B (g)	Response to C (g)
0.1 mL of 0.75 μg/mL		0.27		
0.2 mL of 0.75 μg/mL		0.50		0.25
0.4 mL of 0.75 μg/mL		0.85		0.53
0.8 mL of 0.75 μg/mL		1.17	0.15	0.87
0.1 mL of 7.5 μg/mL		1.54	0.42	1.13
0.2 mL of 7.5 μg/mL		1.59	0.63	1.26
0.4 mL of 7.5 μg/mL		1.61	0.94	1.29
0.8 mL of 7.5 μg/mL		1.65	1.10	
0.1 mL of 75 μg/mL			1.03	
0.2 mL of 75 μg/mL				

(a) Calculate the organ bath concentration (μg/mL) of each agonist.

(b) Plot the log dose–response curves for each drug, and label the axes.

(c) Calculate the EC_{50} value of each drug.

(d) Define the term 'pD$_2$ value'?

(e) Calculate the pD$_2$ for each drug (MW of A = 181.7; MW of B = 96.5; MW of C = 205.1).

(f) Place drugs in order of potency.

(g) Place the drugs in order of maximum response.

2. Why is it important to maintain a constant time cycle in isolated tissue organ bath experiments? What time cycle did you use in your experiments with guinea pig ileum during this semester? [10 marks]

3. You are asked to calculate the concentration of a solution of ACh of unknown concentration using a bioassay. Giving your reasons, which method would you recommend? [10 marks]

4. In a bioassay experiment using a 20 mL organ bath, 0.2 mL of drug A of unknown concentration gave a response of 1.2 g, whist 0.4 mL

of 10 µg/mL of drug A gave a response of 1.4 g and 0.2 mL of a 10 µg/mL solution of drug A gave a response of 0.9 g. What was the concentration of the unknown? [10 marks]

5. (a) How would you make 10 mL of a 1:25 dilution of a drug?
   (b) You require a 1.5 µM solution of acetylcholine chloride (MWt 181.7). How much do you need to weigh out to prepare 10 mL of this solution?
   (c) If you buy 10 mg of a drug (MWt 203.1), what volume of solvent do you need to add to make 10 mM solution. What was the concentration in the acetylcholine solution being assayed?
   (d) What are the advantages of a bracketing assay over a single-point assay?
   (e) How would you carry out a bracketing assay to provide an estimate of the concentration within a margin of error, and minimize the effects of biological variation?

6. With reference to an experiment using guinea pig isolated ileum in an organ bath, the effects of two antagonists, atropine and hexamethonium, on a series were tested on responses to several agonists.

   (a) Complete Table 4.6 indicating whether you would expect inhibition of the agonist a response (+), or no response (−):

Table 4.6 In a guinea-pig isolated ileum, indicate whether there is a positive (+) or negative (−) response of the agonists to the antagonists shown in this table.

	Antagonist	
Agonist	10 µM Hexamethonium	0.1 µM Atropine
Nicotine		
Methacholine		
Acetylcholine		
Carbachol		
Histamine		

   (b) What particular problems are associated with studying nicotine responses in the guinea pig ileum?
   (c) From these results, draw a diagram of the parasympathetic innervation of the guinea pig ileum indicating the location of the receptor types at which each of the agonists and antagonists act.

7. In an experiment using a guinea pig isolated ileum to determine the pA_2 of diphenhydramine the following data was obtained (Table 4.7).

Table 4.7   Table for concentration-response data in guinea-pig isolated ileum.

[Histamine] (μg/mL)	Log[histamine] (μg/mL)	Responses (mm) in the presence or absence of diphenhydramine				
		0 μg/mL	3.8 μg/mL	6.0 μg/mL	17.3 μg/L	45.4 μg/mL
20	1.301	8				
40	1.602	34				
50	1.699	61	8			
80	1.903	90	21	7		
100	2.000	125	43	8	4	
200	2.301	131	105	58	18	
400	2.602	129	129	120	91	9
600	2.778		130		120	41
800	2.903				141	58
1000	3.000				140	78
1200	3.079					89
2000	3.301					125
4000	3.602					127

(a) Plot the five dose–response curves for histamine in the presence or absence of diphenhydramine.

(b) From these curves, calculate the dose–shift ratios (X) at each concentration of antagonist. First calculate the concentration of histamine (D) required to produce the same response from each of the five curves to obtain D_0, D_1, D_2, D_3 and D_4.

(c) Complete Table 4.8, given that the molecular weight of diphenhydramine hydrochloride is 291.8.

Table 4.8   Complete theis table in preparation for the construction of a Schild plot.

[Diphen] μg/mL	[Diphen] M	−Log[diphen] M	Concentration Ratio (X)	(X − 1)	Log(X − 1)

(d) Construct a Schild plot by drawing a graph of −log[diphenhydramine] M (x-axis) against log(X − 1).

(e) From this graph, calculate (a) pA_2 and (b) the slope. Do pA_x values have any units?

(f) Does this data suggest that diphenhydramine is acting as an antagonist of weak, moderate or high potency? Is diphenhydramine acting competitively in this experiment?

ANSWERS TO PROBLEMS

1. Table 4.5 has been completed in Table 4.9

Table 4.9 Completed table from Table 4.5 in Question 1.

Addition to organ bath	[Agonist] μg/mL	Response to A (g)	Response to B (g)	Response to C (g)
0.1 mL of 0.75 μg/mL	$0.1 \times 750/15 = 5$	0.27		
0.2 mL of 0.75 μg/mL	10	0.50		0.25
0.4 mL of 0.75 μg/mL	20	0.85		0.53
0.8 mL of 0.75 μg/mL	40	1.17	0.15	0.87
0.1 mL of 7.5 μg/mL	50	1.54	0.42	1.13
0.2 mL of 7.5 μg/mL	100	1.59	0.63	1.26
0.4 mL of 7.5 μg/mL	200	1.61	0.94	1.29
0.8 mL of 7.5 μg/mL	400	1.65	1.10	
0.1 mL of 75 μg/mL	500		1.03	
0.2 mL of 75 μg/mL	1000			

(a) Calculate the organ bath concentration (μg/mL) of each agonist.
(b) Plot the log dose–response curves for each drug, and label the axes (Figure 4.9).

Answer:
(c) Calculate the EC_{50} value of each drug.

Answer: A = 18.36, B = 80.67 and C = 24.9 μg/mL.
(d) Define the term 'pD$_2$ value'?

Answer: pD$_2$ is the negative log of molar concentration that processes half the maximum response = $-\log EC_{50}$ (M).

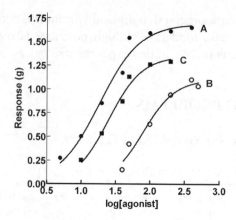

Figure 4.9 Log concentration–response curves for the three cholinesters, A, B and C.

(e) Calculate the pD_2 for each drug (Mr A = 181.7; Mr B = 96.5; Mr C = 205.1).

Answer: A = 7, B = 6.08, C = 6.92.

(f) Place drugs in order of potency.

Answer: A>C>>B.

(g) Place the drugs in order of maximum response.

Answer: A>C>B.

2. Why is it important to maintain a constant time cycle in isolated tissue organ bath experiments? What is the recommended time cycle for this experiment?

Answer: A constant time cycle is important in order to obtain reproducible responses.

The time cycle is 30 s contact, 90 s washout, total 2 min time cycle.

3. You are asked to calculate the concentration of a solution of ACh of unknown concentration using a bioassay. Giving your reasons, which method would you recommend?

Answer: Bracketing assay. See Section 2.2.2.

4. In a bioassay experiment using guinea pig isolated ileum in a 20 mL organ bath, 0.2 mL of drug A of unknown concentration gave a response of 1.2 g, whilst 0.4 mL of 10 µg/mL of drug A gave a response of 1.4 g and 0.2 mL of a 10 µg/mL solution of drug A gave a response of 0.9 g. What was the concentration of the unknown?

Answer:

Organ bath concentration	Response (g)
$0.4 \times 10 \times 1000/20 = 200$ μg/mL	1.4
$0.2 \times 10 \times 1000/20 = 100$ μg/mL	0.9

0.2 mL of unknown produced [organ bath] of antilog 2.18 = 151 μg/mL.

So there was $151 \times 20 = 3020$ ng $= 3.02$ mg in the bath and in 0.2 mL unknown.

So 1mL unknown contains $5 \times 3.02 = 15.1$ mg/mL.

5. (a) How would you make 10 mL of a 1:25 dilution of a drug?

   Answer: $10/25 = 0.4$, so add 0.4 mL of drug solution +9.6 mL water to make 10 mL.

   (b) You require a 1.5 μM solution of acetylcholine chloride (MWt 181.7). How much do you need to weigh out to prepare 10 mL of this solution?

   Answer: 1.5 μM = 1.5 μmol/L = 15 nmol/10 mL.
   1 nmol ACh = 181.7 ng, so 15 nmol = $181.7 \times 15 = 2725$ng = 2.7 μg.

   (c) If you buy 10 mg of a drug (MWt 203.1), what volume of solvent do you need to add to make 10 mM solution?

   Answer: Calculate how many μmol you bought = 10 000 μg/ 203.1 = 49.23 μmol. We need 10 mM (= 10 mmol/L or 10 μmol/mL). So dissolve your 49.23 μmol in 4.92 mL to obtain 10 μmol/mL (10 mM).

6. (a) See Table 4.10

   Table 4.10   Completed table for question 6.

Agonist	Antagonist	
	10 μM Hexamethonium	0.1 μM Atropine
Nicotine	−	−
Methacholine	+	−
Acetylcholine	+	−
Carbachol	+	−
Histamine	+	+

   (b) What particular problems are associated with studying nicotine responses in the guinea pig ileum?

   Answer: A decrease in response is seen after rapidly repeated applications of nicotine which is termed tachyphylaxis.

(c) From these results, draw a diagram of the parasympathetic innervation of the guinea pig ileum indicating the location of the receptor types at which each of the agonists and antagonists act.

7. The inhibition of histamine responses by diphenhydramine is shown
in Table 4.11

Table 4.11   Inhibition of histamine responses by diphenhydramine.

[Histamine] (µg/mL)	Log[histamine] (µg/mL)	Responses (mm) in the presence or absence of diphenhydramine				
		0 µg/mL	3.8 µg/mL	6.0 µg/mL	17.3 µg/L	45.4 µg/mL
20	1.301	8				
40	1.602	34				
50	1.699	61	8			
80	1.903	90	21	7		
100	2.000	125	43	8	4	
200	2.301	131	105	58	18	
400	2.602	129	129	120	91	9
600	2.778		130		120	41
800	2.903				141	58
1000	3.000				140	78
1200	3.079					89
2000	3.301					125
4000	3.602					127

1. Plot the five dose–response curves for histamine in the presence or
absence of diphenhydramine (Table 4.10):

Answer: (see Figure 4.10)

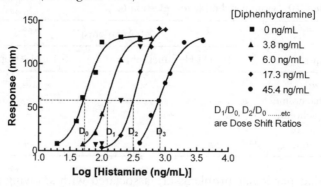

Figure 4.10   Concentration–response curves for histamine in the presence and
absence of diphenhydramine.

2. From these curves, calculate the dose–shift ratios (X) at each
concentration of antagonist. First calculate the concentration of

histamine (D) required to produce the same response from each of the five curves to obtain D_0, D_1, D_2, D_3 and D_4. Complete Table 4.12.

Table 4.12 Completed table for the calculation of the concentration ratio.

[Diphen] μg/mL	Log[C] μg/mL	[C] μg/mL	CR = C/C$_0$
0	1.74	$C_0 = 55.0$	–
3.8	2.09	$C_1 = 123.0$	2.24
6.0	2.23	$C_2 = 177$	3.23
17.3	2.44	$C_3 = 278$	5.07
45.4	2.91	$C_4 = 796$	14.49

3. Complete Table 4.13, given that the molecular weight of diphenhydramine hydrochloride is 291.8.

Answer:

Table 4.13 Completed table for the construction of a Schild plot.

[Diphen] μg/mL	[Diphen] M	−Log[diphen] M	CR	(CR − 1)	Log(CR − 1)
3.8	1.30×10^{-8}	7.88	2.24	1.24	0.09
6.0	2.09×10^{-8}	7.68	3.23	2.23	0.35
17.3	5.92×10^{-8}	7.22	5.07	4.07	0.61
45.4	1.57×10^{-7}	6.80	14.49	13.49	1.13

4. Construct a Schild plot by drawing a graph of $-\log[\text{diphenhydramine}]$ M (x-axis) against $\log(X - 1)$. See Figure 4.11.

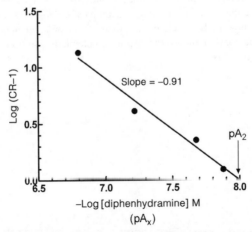

Figure 4.11 Schild plot for the antagonism of histamine responses in the guinea pig isolated ileum.

5. From this graph, calculate (a) pA_2, (b) pA_{10} and (c) the slope. Do pA_x values have any units?

    Answer: No.

    $pA_2 = 7.99$

    Slope $= -0.91$

6. Does this data suggest that diphenhydramine is acting as a competitive antagonist in this experiment?

    Answer: Yes, slope is very close to -1.

7. Does this data suggest that diphenhydramine is acting as an antagonist of weak, moderate or high potency?

    Answer: Moderate potency (<7 is low, and >8 is high potency).

REFERENCES

Ahlquist, R.P. (1948) A study of the adrenotropic receptors. *Am. J. Physiol.* 153(3): 586–600.

Arunlakshana, O. and Schild, H.O. (1959) Some quantitative uses of drug antagonists. *Brit. J. Pharmacol.* 14: 48–58.

Rang, H.P., Ritter, J., Flower, R. and Henderson, G. (2012) *Rang and Dale's Pharmacology*, 7th edn. Elsevier Life Sciences.

Spedding, M. (1982) Assessment of "Ca^{2+}-antagonist" effects of drugs in K^+-depolarized smooth muscle. Differentiation of antagonist sub-groups. *Naunyn Schmiedebergs Arch. Pharmacol.* 318: 234–240.

Finkleman, B. (1930) On the nature of inhibition in the intestine. *J. Physiol. (Lond.)* 70: 531–540.

Gaddum, J.H. (1937) The quantitative effects of antagonistic drugs. *J. Physiol.* 89: 7P–9P.

Magnus, R. (1904) Versuche am überlebenden dünndarm von säugethieren I. Mittheilung. ("Experiments on the surviving intestine of mammals"). *Pflugers. Arch.* 102: 123–151.

Waterfield, A.A., Smokum, R.W., Hughes, J., Kosterlitz, H.W. and Henderson, G. (1977) In vitro pharmacology of the opioid peptides, enkephalins and endorphins. *Eur. J. Pharmacol.* 43: 107–116.

5

Cardiovascular Preparations

Because cardiovascular diseases have been such a common health risk, it has been a priority to understand the pharmacology of the cardiovascular system in order to develop new drugs. Due to the complexity of the various tissues and cell types involved numerous preparations have been devised. One of the earliest was the Langendorff isolated perfused heart preparation. Once removed from the animal the preparation provides a useful isolated model of the heart, but suffers from the drawback that that it is not perfused in a physiological manner. It is often referred to as retrograde perfusion as Ringer solution enters the aorta and leaves through the venous vessels. However, pacemaker cells will continue to generate cardiac action potentials which are conducted from the atria to the ventricles as *in vivo*, and the heart will continue to beat for many hours. An improved heart perfusion technique is the working isolated heart preparation which can develop pressure equivalent to arterial blood pressure. Whilst the perfused heart is a good model in which to study cardiac pharmacology, it requires some skill and practice to set up and also requires specialized equipment. A simpler heart preparation suitable for student use is the isolated auricle preparation. The same organ bath equipment as used for the isolated guinea pig ileum can be used. Obviously isolated heart preparations contain a large variety of cell types, and the exact cell type on which drug receptors are located may not be unequivocally known. A more advanced technique, which is especially useful for studying the electrophysiology and the actions of drugs on pacemaker potentials, is provided by isolated cardiomyocytes. These can be grown in tissue culture, where they show the remarkable property of assuming a co-ordinated, rhythmic beating in a tissue culture

Practical Pharmacology for the Pharmaceutical Sciences, First Edition. D. Michael Salmon.
© 2014 John Wiley & Sons, Ltd. Published 2014 by John Wiley & Sons, Ltd.

dish. They respond to noradrenaline and acetylcholine by altering the frequency of action potentials and chronicity of the beating of the cells, and provide a unique preparation in which to investigate the molecular mechanisms of action of these neurotransmitters and drugs that control heart rate.

Several vascular preparations have been used, but many require skill and practice to set up. Of the various vascular preparations that have been used, one of the relatively easy ones to set up is the isolated thoracic aorta preparation. The aorta from several species have been used, but it is obviously easier using larger animals. It is possible to use the aorta from the rat or guinea pig quite satisfactorily. Famously, this preparation played an important role in establishing the role of nitric oxide in the cardiovascular system.

5.1   ISOLATED PERFUSED HEART PREPARATIONS

The isolated, perfused heart is a well-established preparation to demonstrate the actions of a wide variety of classes of drugs such as sympathomimetics, parasympathomimetics and histamine. The actions of arrhythmics and anti-arrhythmics can also be demonstrated. Cardiac glycosides, such as ouabain, have a strong inotropic effect with little effect on rate. At toxic doses the heart becomes markedly arrhythmic and stops in systole. Sympathomimetic drugs produce both inotropic and chronotropic responses, which are mediated mainly via β_1-receptors, although a small population of β_2 and dopamine receptors are also found in the heart. Parasympathetic responses occur via M_3 cholinoceptors which produce negative chronotropic responses, and acetylcholine in large doses will stop the isolated heart completely for a short time. Nicotine has interesting effects on the isolated heart. It elicits a transient decrease in rate and force, followed by a larger increase in inotropy and chronotropy. The first phase is antagonized by atropine and α-bungarotoxin, but not by hexamethonium, whilst the second phase is inhibited by the β-blocker, timolol and hexamethonium. This is an evidence for two types of nicotinic receptors in the heart (Ji *et al.*, 2002).

For many years, isolated heart preparations from small mammals have proved to be invaluable models in which to study the pharmacology of the heart (reviewed by Sutherland and Hearst, 2000). However, it must be recognized that none of the available variations of the isolated preparation conserve all the physiological controls present *in vivo*. Because none of the isolated preparations are innervated, they lack all intrinsic autonomic control; and the basal heart rate is usually determined by the rate of the pacemaker cells in the right atrium. The two major types of isolated heart preparation are the classic Langendorff 'retrograde perfusion' preparation (Langendorff, 1895). The term 'retrograde' is used since the perfusion fluid enters the aorta close to the heart and passes through the cardiac vascular system in the opposite direction to that encountered *in vivo*. The more complex and technically demanding working heart preparation was first described by Neely *et al.* (1967). Here the perfusion passes through the coronary vessels in normal physiological direction. The heart is required to perform work in pumping the perfusion fluid to a height of 60–100 cm of water, equivalent of normal arterial pressure. The preparation can be perfused with buffer or blood, but performs best with the superior oxygen delivery allowed with blood. The working heart preparation is particularly suited to the investigation of ischaemia, ventricular fibrillation and arrhythmias.

5.1.1   The Langendorff Preparation

Experimental Conditions

Organ bath	Open water-jacketed chamber, 100 mL
Ringer solution	McEwans or Krebs (filtered)
Aeration	95% O_2/5% CO_2
Bath temperature	37°C
Transducer	Isometric
Resting tension	3–4 g
Equilibration period	30–45 min

This involves the cannulation of the aorta which is then attached to a reservoir containing oxygenated perfusion fluid. This fluid is then delivered in a retrograde direction down the aorta either at a constant flow rate (delivered by an infusion or roller pump) or at a constant hydrostatic pressure (usually in the range of 60–100 mm Hg). In both instances, the aortic valves are forced shut and the perfusion fluid is directed into the coronary ostia thereby perfusing the entire ventricular mass of the heart, draining into the right atrium via the coronary sinus.

There are several variations of the equipment used for this preparation, but for a laboratory demonstration, simple equipment consisting of a glass or plastic reservoir for the Krebs' solution, a warming coil utilizing a condenser and a heated organ bath with a hole at the bottom is satisfactory. To monitor cardiac contractions and heart rate, the simplest method is to attach an isometric transducer to the apex of the heart. This is achieved using a heart clip tied to a thin cotton thread which passes round a pulley arrangement to the transducer. A more sophisticated technique involves the insertion of an inflatable intraventricular balloon. For most mammalian heart preparations, Krebs–Heinseleit solution can be used, but one important precaution is to ensure that it has been filtered to remove any fine particulate material. It is usual to pre-treat the animal with heparin to prevent blood clots and remove the heart under anaesthesia. However, since administration of any substance to an animal is a licensed procedure, it is possible to avoid these procedures by removing the heart as rapidly as possible (about 30 s) from a deceased animal and ensuring that cannulation of the aorta is performed as rapidly as possible. The guinea pig is killed by cervical dislocation. Once the animal is anaesthetized the heart can be excised. Generally, the diaphragm is accessed by a transabdominal incision and cut carefully to expose the thoracic cavity. The thorax is opened by a bilateral incision along the lower margin of the last to first ribs, the thoracic cage is then reflected over the animals head, exposing the heart. Some investigators then cradle the heart between their fingers (it is essential to do this gently to avoid muscle damage) and then lift the heart slightly before incising the aorta, vena cava and pulmonary vessels. The heart is rapidly exposed, excised and plunged into ice-cold Krebs–Ringer solution to stop the heart beating to minimize to limit any ischaemic injury during the period between excision and the restoration of vascular perfusion and metabolic demand. The aorta is located and cut to within 1 cm of the heart. A cannula made of glass, plastic or thin-walled stainless steel is used. The external diameter is typically similar to, or slightly larger than, that of the aorta (about 3 mm for a heart from a 250 g rat). Several small circumferential grooves is usually machined into the distal end of the cannula to prevent the aorta from slipping off. Some cannulae are heated with water jackets to prevent any unwanted fall in perfusate placed about 0.5 cm into the aorta, and secured firmly using cotton ties. A water-jacketed reservoir, situated above the aortic cannula, contains the perfusion fluid which is oxygenated via a sintered glass gas distributor (for bicarbonate-based perfusion fluids, $95\%O_2 + 5\%CO_2$ is normally used). It is advisable to have the perfusion fluid gently dripping from the aortic cannula prior to cannulation since this helps minimize the chance of air emboli at the time the heart is attached to the cannula. Cannulation is aided by cutting along the aortic arch to open it, thus giving a larger area for cannulation. Hearts should be held gently between the tips of blunt-ended fine curved forceps, taking care to avoid stretching or ripping of the aortic wall. The aorta is then gently eased over the end of the cannula, taking care not to insert the cannula too far into the aorta since this would occlude the coronary sinus orifice or damage the aortic valve. The cannula is clamped to the aorta using a small, blunt-nosed artery clip, whilst a ligature is rapidly tied around the aorta, locking into the grooves; the artery clip can now be removed. In the case of the flanged cannula the aorta is slid down the cannula so that the tie is against the flange. Full flow of perfusate should be initiated as soon as the heart is mounted on the cannula. Ringer solution enters the aorta in a retrograde direction, which closes the aortic valve, and fluid is directed through the coronary circulation. Drainage of coronary perfusate from the right side of the heart via the pulmonary artery should be unimpeded, however, in the course of cannulation it is possible to accidentally ligate the pulmonary artery. Thus, to facilitate adequate drainage it is advisable to make a small incision in the base of the pulmonary artery using small pointed scissors. The tap from the reservoir is opened to exert a hydrostatic pressure of about 70 cm. The heart quickly warms to 37°C and starts to beat. A heart clip with a length of thread is attached to the tip of the ventricles. The cotton thread is fed through the pulley system and attached to the isometric transducer. The recorder is switched on and traces of heart force and rate are obtained. Two-channel recording equipment is required to record force (g) and heart rate (beats per minute). The preparation should be allowed to equilibrate for 30–60 min to allow a steady force and heart rate to be established. Drugs are introduced into the aorta (vol. 0.2 mL or less) using a syringe and fine bore catheter tubing (18 G, 1.27 mm external diameter) inserted through a rubber diaphragm on one arm of the three-way adapter. Before addition of each drug the previous drug solution must be expelled from the tubing by injection of Krebs solution.

The full range of drugs affecting heart function can be tested using this preparation. Thus the actions of autonomic neurotransmitters on rate and force can be explored, and the types of cholinoceptors and adrenoceptors demonstrated. It is wise to establish that these responses are obtained in a preparation before proceeding to test novel responses to drugs.

5.1.2   Cardiac Interactions of Anti-asthma Drugs

The experiment described here is intended to demonstrate the potentially dangerous effect on the heart if two anti-asthma drugs are administered together. The β_2-agonist, salbutamol is well known as a potent bronchodilator and at therapeutic doses has weak sympathomimetic effects on the heart. Theophylline is another bronchodilator drug, and is prescribed for some asthmatic patients. It is a methylxanthine that acts as a cyclic AMP (cAMP) phosphodiesterase inhibitor, as well as being a non-selective adenosine receptor antagonist. Both these actions will increase intracellular cAMP concentrations in myocytes. Since activation of β_1-adrenoceptors at high doses of salbutamol also act to increase cAMP concentrations, threshold doses of theophylline and salbutamol will be synergistic and produce tachycardia and increased inotropy. Therefore, it is strongly advised not to use these two drugs together if potentially fatal tachycardia is to be avoided. Here the potency of the specific β_1-agonist, isoproterenol, is first compared with that of salbutamol to demonstrate the low potency of salbutamol. A dose of theophylline that produces a very small response is administered, followed by a threshold dose of salbutamol. When theophylline is administered followed by salbutamol without any washout, large inotropic and chronotropic responses are found.

Protocol

The following stock solutions should be available: make 5 serial (1:10) dilutions (in 0.9% NaCl) of 10 mg/mL isoprenaline (isoproterenol) hydrochloride containing 1 mg ascorbate, 5 serial dilutions (in 0.9% NaCl) of 10 mg/mL salbutamol sulphate and a 10 mg/mL theophylline (dissolve in 0.1 M Na_2CO_3).

1. Using a 1 mL syringe attached to the side-arm of the three-way tap via the fine catheter tubing, slowly inject 0.2 mL of 1 µg/mL salbutamol. No response should be observed at this low dose. Wait 5 min, then increase the dose ten times. Repeat this procedure until maximum response for rate and force are recorded. If time allows, the dose can be increased in steps of 5-fold instead of 10-fold.
2. Inject 0.1 mL of 0.9% saline to clear the catheter tubing of salbutamol. Check that the starting baseline tension has been restored.

3. Inject 0.1 mL of 0.1 μg/mL isoprenaline. Observe the responses for 5 min, then again increase the dose by 10 (or if time allows, five) times as previously. Continue with this procedure until maximum responses are recorded. Inject 0.9% NaCl to ensure that the isoprenaline has been removed from the fine cannula injection tubing.
4. Select a threshold dose of salbutamol at the bottom of the dose–response curve. This will be in the region of 0.2 mL of 10 μg/mL of salbutamol. Wait 5 min, then wash out with 0.2 mL of 0.9% saline.
5. Inject 0.2 mL of 10 mg/mL theophylline, wait 5 min, then wash out as above. No response should be observed. Wash out the theophylline with 0.9% NaCl.
6. Inject 0.2 mL of 1 mg/mL theophylline, then without washing out inject 0.2 mL of 10 μg/mL salbutamol.

A typical example of the Chart® recording obtained for this experiment is shown in Figure 5.1.

It should be seen that salbutamol is about 100 times less potent than isoprenaline, suggesting that both drugs were acting on β_1-adrenoceptors in the heart. At the doses selected, neither salbutamol nor theophylline produce responses, but together they produce a clear response.

Questions

1. Measure and tabulate changes in force (g) and heart rate (bpm).
2. Draw a graph of log dose (ng) against force and heart rate. A typical example of the graph obtained is shown in Figure 5.2.
3. Estimate relative potency (as dose ratio) for salbutamol and isoprenaline.
4. Draw a histogram of salbutamol, theophylline and salbutamol with theophylline.
5. Explain these results and their possible clinical significance.

5.1.3 The Rat Isolated Auricle Preparation

This preparation is adapted from that described by Saunders and Thornhill (1985). An isometric transducer is connected to a bridge amplifier and an ADInstruments PowerLab amplifier and AD converter. Responses are recorded by a computer running Chart software.

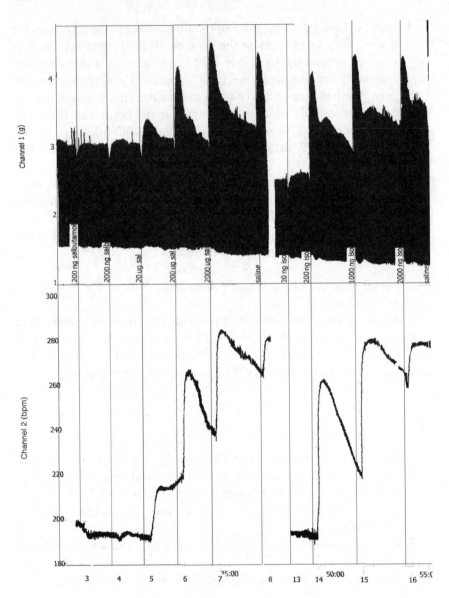

Figure 5.1 A Chart recording of the responses of guinea pig Langendorff heart preparation to isoprenaline and salbutamol.

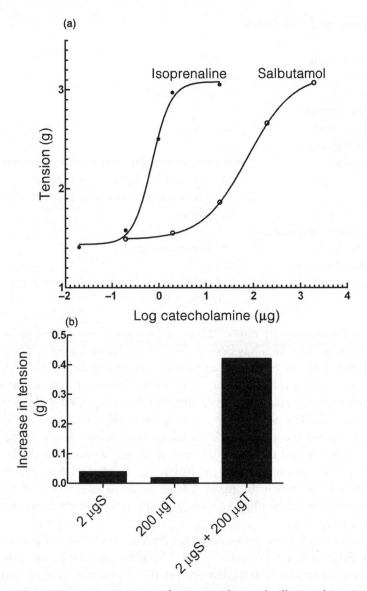

Figure 5.2 (a) Dose–response curves for isoprenaline and salbutamol in a Langendorff preparation of a guinea pig heart. The relative potencies of the two agonists can be measured from the third graph. (b) Bar graphs of the responses to salbutamol, theophylline and theophylline and salbutamol together in the guinea pig Langendorff heart preparation.

Experimental Conditions

Organ bath	20 mL
Ringer	Krebs
Aeration	95% CO_2/5% O_2
Temperature	37°C
Transducer	Isometric
Resting tension	8 g
Dosing	Obtain cumulative concentration–response curves as the preparation may take 5–10 min to recover a steady rate after washing.
Electrical stimulation	
Voltage	5 V above maximal
Pulse duration	20 ms
Frequency	6 Hz (mimics resting heart rate of a rat, 360 bpm)

The heart is rapidly removed from a guinea pig and placed in Locke–Ringer solution which has been cooled to 4°C in an ice bath. This stops the beating of the heart and minimizes damage to heart tissue due to ischaemia during the dissection of the heart. The ventricles are cut away and surrounding tissue removed until all that remains are the left and right auricles. The auricles are appendages of the atria. A cotton thread is tied around the tip of each auricle, one of which is anchored to a metal support and the other left free to attach to the transducer. The tissue is then mounted in an organ bath and the thread is attached to the transducer. When the tissue warms up to 37°C it will start to beat again, and should be allowed to equilibrate when the rhythmic beating becomes regular.

A concentration–response curve can be obtained at organ bath concentrations between 10^{-8} M and 5×10^{-6} M of noradrenaline. The presence of β_1-adrenoceptors can be demonstrated by obtaining a concentration–response curve with isoprenaline (a non-selective β-adrenoceptor agonist) between similar concentrations. This should be inhibited by 10^{-6} M atenolol (a β_1-selective adrenoceptor antagonist).

5.2   THORACIC AORTA PREPARATION

It is generally thought that the innervation of blood vessels is mainly by the sympathetic division of the autonomic nervous system. In the aorta, adrenoceptors mediating constriction are mainly of the α_1 sub-type. β_2-receptors mediate dilation, but mainly in skin and peripheral tissues. Only some blood vessels have parasympathetic innervation, but acetylcholine causes vasodilation *in vivo*. In isolated aortic strip preparations, acetylcholine was notorious for producing erratic responses, sometimes causing contraction, sometimes dilation. This problem was resolved in the seminal report by Furchgott and Zawadzki (1980), who found that if great care was taken not to damage the inner endothelial lining of the aorta, acetylcholine reproducibly produced relaxation. A similar finding is found for some other vasodilators, such as bradykinin and substance P. The isolated aortic strip preparation was instrumental in elucidating the essential role of the endothelium in controlling vascular tone. The factor released by endothelial cells which subsequently acted on vascular smooth muscle was termed endothelium-derived relaxing factor (EDRF). Subsequently, it was demonstrated that this relaxing factor was in fact the unstable free radical, nitric oxide (reviewed by Furchgott, 1998 in his Nobel lecture). Some vasodilators are not endothelium dependent, and a number of them are used in the treatment of hypertension. Glyceryl trinitrate (and other nitric oxide donors) act directly on (more accurately in) smooth muscle cells to release nitric oxide which activates guanylate cyclase. Other vasodilators, such as calcium channel antagonists (e.g. nifedipine), act by antagonizing L-type calcium channels (see Section 4.2.6).

Experimental Conditions

Organ bath volume	20 mL
Ringer solution	Krebs
Aeration	95% O_2/5% CO_2
Bath temperature	37°C
Transducer	Isometric
Resting tension	2 g
Contact time	3–5 min
Dose cycle	Obtain cumulative curves. Allow 5–10 min to relax.

Method

A short section is carefully removed from the rat and dissected free from connective tissue and fat. Utmost care must be taken in handling the aorta section so as not to damage the interior endothelial lining of the vessel. 1.5–2 cm segments of rat aorta are mounted horizontally in a water-jacketed organ bath of 10 mL filled with Krebs–Heinseleit solution (see Section 2.3) maintained at 37°C. The solution is aerated with a gas mixture containing 95% O_2:5% CO_2. The rings are suspended horizontally on a pair of stainless steel hooks, one of which was fixed to an L-shaped rod inside the chamber and the other to an isometric transducer (PowerLab ML750). The stainless steel hook is connected to the force displacement transducer. Isometric contractions are measured and recorded continuously in a computer by using the Chart software. Arterial rings are equilibrated in Krebs–Henseleit solution for 1 h at 1 g optimum resting force. In some rings, the endothelium can be removed gently by rubbing the luminal surface of the ring with a roughened polyethylene tube. At the end of the equilibration period, viabilities of the arterial segments with and without endothelium are checked by depolarization with KC1 (60 mM) and phenylephrine (10^{-5} M). The response of this tissue is relatively slow, and it will take 5 min or more to reach full contraction. Cumulative dose–response curves are therefore routinely employed. The effectiveness of endothelium removal is confirmed by the inability of acetylcholine (10^{-6} M) to induce relaxation of phenylephrine-precontracted rubbed rings when less than 10% of maximum relaxation should be obtained.

5.2.1   Drugs Regulating Nitric Oxide-mediated Relaxation

This experiment is intended to demonstrate that some vasodilators are dependent on the presence of an intact endothelium and others act independently of the endothelium. Endothelium-dependent vasodilators, such as acetylcholine and bradykinin, act via a receptor on the endothelium membrane. This results in a rise in endothelial intracellular [Ca^{2+}], which activates the enzyme endothelial nitric oxide synthase (eNOS). Nitric oxide produced diffuses into adjacent smooth muscle cells where it activates soluble guanylate cyclase. The increase in cGMP results in a decrease in intracellular [Ca^{2+}] in smooth muscle cells. Here, acetylcholine is used as an endothelium-dependent vasodilator. The idea that acetylcholine-stimulated vasodilation requires activation of eNOS is demonstrated by inhibiting this enzyme with N_ω-nitro-L-arginine methyl ester hydrochloride (L-NAME). If a source of NO is supplied exogenously in the form of a NO donor, such as sodium nitroprusside (SNP) or glyceryl trinitrate, the vasodilation can be restored.

Procedure

The following stock concentrations of drugs should be available: 2 mM phenylephrine containing few milligrams of ascorbic acid, 2 mM acetylcholine, 10 mM L-NAME and 10 mM and 10 µM SNP (protected from light).

1. The tissue is sub-maximally contracted with 1 µM phenylephrine and allowed the response to stabilize.
2. Cumulative concentration–relaxation responses are obtained with acetylcholine (0.1–10 µM). If the tissue fails to relax in response to acetylcholine, this indicates that the endothelium has been damaged whilst setting up the preparation.
3. Without washing out, add 100 µM L-NAME. A slow contraction should be seen over a period of about 10 min.
4. Again without washing out, cumulatively add SNP over the range 10 nM–10 µM. A further relaxation should be seen.

Questions

1. Measure and tabulate the responses.
2. Draw cumulative concentration–response curves for acetylcholine and SNP.
3. Explain how these responses were brought about (a) the mechanism underlying the relaxation effect of acetylcholine, (b) why does L-NAME overcome this effect and (c) how SNP can cause a relaxation, even in the presence of L-NAME.

REFERENCES

Furchgott, R.F. and Zawadzki, J.V. (1980) The obligatory role of endothelial cells in the relaxation of arterial smooth muscle by acetylcholine. *Nature* 288(5789): 373–376.

Furchgott, R.F. (1998) Endothelium-derived relaxing factor: discovery, early studies and identification of nitric oxide. Nobel Prize lecture. Available at http://www.nobelprize.org/nobel_prizes/medicine/laureates/1998/furchgott-lecture.pdf (accessed 9 September 2013).

Ji, S., Tosaka, T., Whitfield, B.H., Katchman, A.N., Kandil, A., Knollmann, B.C. and Ebert, S.N. (2002) Differential rate responses to nicotine in rat heart: evidence for two classes of nicotinic receptors. *J. Pharm. Exp. Ther.* 301: 893–899.

Langendorff, O. (1895) Untersuchungen am uberlebenden Saugethierherzen (Studies on the surviving heart of mammals). *Pflugers Archives fur die Gesamte Physiologie des Menschen and der Tiere.* 61, 291–332.

Neely, J.R., Liebermeister, H., Battersby, E.J. and Morgan, H.E. (1967) Effect of pressure development on oxygen consumption by isolated rat heart. *Am. J. Physiol.* 212: 804–814.

Saunders, W.S. and Thornhill, J.A. (1985) No inotropic action of enkephalins or enkephalin derivatives on electrically-stimulated atria isolated from lean and obese rats. *Br. J. Pharmacol.* 85: 513–522.

Sutherland, F.J. and Hearst, D.J. (2000) The isolated blood and perfusion fluid perfused heart. *Pharmacol. Res.* 41: 613–627.

6

Skeletal Muscle

6.1 TYPES OF SKELETAL MUSCLE

Skeletal muscle is frequently referred to as striated or voluntary muscle, but the first term is the only one that really accurately describes
the wide variety of muscles that fall within this category. Other muscles, notably cardiac muscle, are also striated, whilst skeletal muscle
of the diaphragm is largely under involuntary control. Anatomically,
there are a range of fibre types, including red, white and intermediate.
Functionally, they range from fast acting, which fatigue easily, to slow
acting that are resistant to fatigue. Since many drugs affecting skeletal
muscle act at the neuromuscular junction, pharmacologically the most
important factor is the type of innervation each muscle type receives.
All striated muscles are innervated by thickly myelinated somatic nerves,
which run directly from the central nervous system without intermediate ganglia. The distribution of neuromuscular end plates on the muscle
fibres fundamentally affects the type of response of the muscle to nervous stimulation. Most mammalian muscles have a single end plate on
the muscle surface, and are described as *focally innervated*. An action
potential arriving at the end plate on a muscle fibre causes activation
of nicotinic receptors which results in depolarization of the muscle cell.
This depolarization rapidly spreads across tight junctions between cells
and is transmitted throughout the fibre. There is a co-ordinated contraction of the muscle seen as an all-or-nothing response. Increasing the
frequency of action potential will cause an increase in the frequency
of contractions. In contrast, nerves supplying skeletal muscles of many

Practical Pharmacology for the Pharmaceutical Sciences, First Edition. D. Michael Salmon.
© 2014 John Wiley & Sons, Ltd. Published 2014 by John Wiley & Sons, Ltd.

amphibia and birds terminate in very many end plates on each muscle fibre, and are described as *multiply innervated*. Action potentials arriving at these end plates cause only a local depolarization, and result in a graded contraction, or contracture. As action potentials arrive at the end plate with increased frequency, more muscle cells become depolarized and this is seen as a graded increase in contracture. By analogy with electromagnetic signals, the difference in the response of focally innervated and multiply-innervated muscles could be described as digital or analog respectively. A major difference between *in vitro* preparations of focally innervated and multiply-innervated muscles is that focally innervated muscles, such as the rat diaphragm or frog gastrocnemius leg muscle, do not respond when acetylcholine is added to the bath. Multiply-innervated muscles do respond, and produce a graded contracture which increases as the bath concentration of the nicotinic receptor agonist is increased. Consequently, it is far easier for students to investigate the action of drugs at the neuromuscular junction using a multiply-innervated preparation, although the type of response will not be the same as in most mammalian skeletal muscles. Thus suxamethonium is used to cause short-term paralysis of mammalian skeletal muscles, in fact causes contraction in a multiply-innervated muscle such as the frog rectus abdominis. Experiments on two multiply-innervated muscles are described here.

6.2   MULTIPLY-INNERVATED SKELETAL MUSCLE PREPARATIONS

Until recently, the frog rectus abdominis was the preparation of choice for student laboratory practicals, since it shows a relatively rapid response or contracture. However, in the past few years it has been difficult to source frogs suitable for this purpose. This may well be due to the reported worldwide decrease in the populations of several amphibian species and this has become a topical conservation issue. An alternative multiply-innervated skeletal muscle preparation is the dorsal muscle of the leech. In the presence of an acetylcholinesterase inhibitor, this is the most sensitive known preparation to acetylcholine (down to 50 pmol added to an organ bath) and is classically used to detect release of acetylcholine released by nerve stimulation. It was therefore instrumental in establishing acetylcholine as a neurotransmitter (Dale and Feldberg, 1934).

6.2.1   Agonists and Antagonists Acting on the Frog Rectus Abdominis

Although the frog rectus abdominis muscle contains mainly multiply-innervated muscle cells, it also contains a few focally innervated fibres. It behaves pharmacologically as a multiply-innervated muscle, producing contractures in response to external agonists. This preparation is an easy one to set up and demonstrate the presence of nicotinic receptors by the actions of agonists and antagonists. In this experiment, the action of ACh, MCh, CCh and nicotine are examined. Although suxamethonium is an antagonist in focally innervated skeletal muscle (e.g. rat diaphragm muscle), it causes a slow contracture in frog rectus abdominis. The response clearly differs from that of acetylcholine, and causes contracture by producing a prolonged depolarization. The contractures produced by acetylcholine are blocked by the competitive nicotinic receptor antagonist, tubocurarine. This is the active principle of the skeletal muscle paralysing poison, curare. Tubocurarine acts as a non-depolarizing blocker.

Method

Experimental Conditions

Organ bath volume	20 mL
Physiological buffer	Frog Ringer
Temperature	Room temperature (20°–22°C)
Aeration	Air
Transducer	Isotonic
Resting tension	0.5 g and a 1 g stretching weight
Dose cycle	~5 min (depends on rate of relaxation)
Contact time	1 min

A small frog weighing about 20 g is stunned, decapitated and the nerves in the spinal column destroyed by inserting a syringe needle (pithing). This is necessary to prevent involuntary reflex contractions. Test for lack of any reflexes by pinching the feet. It is placed on a cork board, and an incision is made in the loose skin extending from the sternum to the bottom of the abdomen. This reveals the two large rectus abdominis muscles. These muscles are excised in one piece. It is divided longitudinally and placed in frog Ringer solution. The muscles from one frog will provide two isolated muscle preparations. A loop of cotton is threaded through one end of the muscle, which is then attached to a metal muscle support rod. A cotton thread is tied to the other end of the muscle and the muscle mounted in the organ bath containing aerated frog Ringer solution. The loose cotton thread is now attached to one arm of an isotonic transducer. To the other arm of the transducer is attached a relaxing counterweight. This consists of a short cotton thread attached to one arm of the transducer, and at the other end is fixed a 1 g counterweight of plasticine. During contractures, the counterweight is removed by placing it on a convenient clamp, and only released and allowed to hang from the arm of the transducer during relaxation periods. Skeletal muscle does not spontaneously relax like smooth muscle, and requires the help of a counter muscle (as in the biceps–triceps arrangement in the arm). With the counterweight removed, arrange that there is about 0.5 g of resting tension and allow a baseline to be established. This may take up to 30 min. A suitable time cycle of events is as follows:

0 min	Raise counterweight, wait for baseline to be re-established.
2 min	Add acetylcholine.
3.5 min	Wash out organ bath and lower counterweight.
~5 min	When a stable baseline is established, raise counterweight.

This time cycle is suitable for acetylcholine, which causes a relatively rapid contracture and relaxation, but other drugs are slower in onset and relaxation takes longer. A slight lengthening of both contact time and washout may be needed. First a sub-maximal response to acetylcholine is obtained with acetylcholine (bath concentration 1 μM). Now test the response to methacholine by producing a bath concentration equal to that which produced a maximal contracture to acetylcholine (~20 μM). No response to methacholine should be found since this cholinester is not an agonist at nicotinic receptors. Now test the response to carbachol, using a bath concentration of about 4 μM. A slower developing contracture which is maintained longer should be found. Similarly, nicotine should produce a clear contracture, but much higher bath concentrations are required (~50 μM). These observations demonstrate that the cholinergic receptor in the rectus abdominis is of the nicotinic sub-type.

The second part of the experiment is to investigate the action of d-tubocurarine (d-TC). Establish a concentration–response curve for acetylcholine, and this is best done as a cumulative curve. To do this, add a low concentration of acetylcholine (bath concentration 1 μM; 50 μL of a 0.2 mM stock solution to a 20 mL organ bath), wait for 90 s contact time, and then, without washing out add another 50 μL to double the bath concentration. After a further 90 s, add 100 μL acetylcholine to double the bath concentration again. Continue in this cycle until a maximum contracture is reached. Enable relaxation by washing out the bath three times and lowering the counterweight. 1 μM d-TC is then added to the organ bath and allowed to equilibrate for 15 min. The concentration–response curve to acetylcholine is then repeated. It will be found that higher concentrations of acetylcholine are now required to obtain equal responses to those found with acetylcholine alone. From the log concentration–response curve it is seen that there is a parallel of the curves suggesting that d-TC is acting as a competitive antagonist. It is possible to construct further concentration–response curves in the presence of higher concentrations of d-TC and construct a Schild plot in order to obtain further confirmation of this idea.

6.2.2 Action of Anticholinesterases on the Dorsal Muscle of the Leech

An alternative, but more difficult preparation, is the dorsal muscle of the leech. These are more plentiful since they are bred for a variety of purposes. The most notable medical use is the extraction of the thrombin inhibitor hirudin for use as an anti-thrombotic agent from the leech species *Hurido medicinalis*. This has largely been superseded by the production hirudin by recombinant DNA technology, but leech farms are still in production. The leech dorsal muscle preparation was developed many years ago and is particularly notable for its extreme sensitivity to acetylcholine, especially in the presence of an anticholinesterase. It appears that the acetylcholinesterase is about ten times more sensitive to physostigmine (serine) than neostigmine. Flacke and Yeoh (1968a, 1968b) extensively characterized the responses and receptors of the leech dorsal muscle.

Like the frog rectus abdominis muscle, the dorsal muscle of the leech is a multiply-innervated skeletal muscle. Such innervation allows a graded response to be obtained after the addition of exogenous agonists, for example, ACh or nicotine. This practical was designed to estimate the pA_2 of a competitive antagonist acting at the neuromuscular junction. Also the effects of eserine (physostigmine) can be examined.

Method

Experimental Conditions

Organ bath volume	20 mL
Physiological buffer	Frog Ringer
Temperature	Room temperature (20°–22°C)
Aeration	Air
Transducer	Isotonic or isometric
Resting tension	2 g
Equilibration time	2 h
Dose cycle	~5 min (depends on rate of relaxation)
Contact time	1 min

Leeches are reared commercially from suppliers such as Biopharm Leeches (UK). They are maintained at a refrigerator with a supply of blood meal (Hirido Mix/Gel from the suppliers). They can be kept in this state for several weeks.

Using an isometric transducer, the tension is set to be about 25 g to stretch the muscle, and the tension gradually declines over the first 1–2 h. When a stable baseline is obtained, the resting tension is decreased to 2 g. The following experiment illustrates the presence of a very high activity of acetylcholinesterase in this tissue. After applying an acetylcholinesterase inhibitor, the potency of acetylcholine increases 500–1000 times (Battacharya and Feldberg, 1958). This is the basis of a sensitive bioassay for acetylcholine.

1. A cumulative dose–response curve to ACh is carried out starting with a bath concentration of 50 µM (10 µL of a 100 mM ACh stock). Leave for a contact time of 2 min. Without washing out, add a further of 10 µL of a 100 mM ACh stock to give a total bath concentration of 100 µM. Double the bath concentration of ACh in this way until the maximum response is reached. Wash out three times and wait for the baseline to return to its previous value.

2. Add 1 mL of 1 mM eserine sulphate to 1 L frog Ringer solution in the reservoir (large 2 L beaker) to give 1 µM eserine. Wait 10 min. Repeat the response curve to ACh, but start with 10 µL of a more dilute stock ACh (100 µM ACh).

3. Add 2 µM d-TC to the organ bath (40 µL of 1 mM d-TC into 20 mL). Wait 10 min, then repeat the ACh cumulative dose–response curve. Wash out three times, and wait for the baseline to be restored.

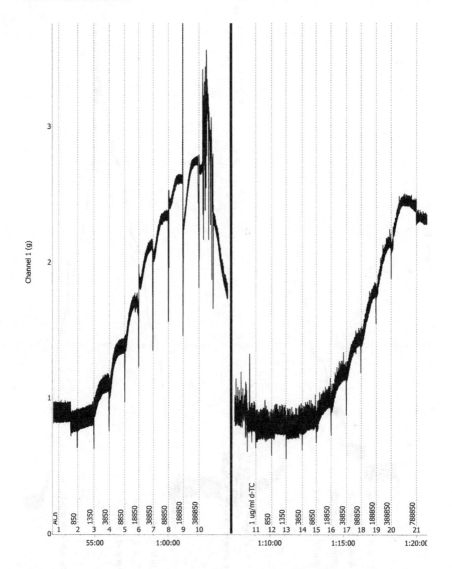

Figure 6.1 Chart® record of cumulative concentration–response curves for acetylcholine in the absence and presence of 1 µM eserine in dorsal muscle of the leech. At each addition, the bath concentration of ACh (nM) is indicated. The abscissa is time (min.).

4. Add 10 µM (200 µL of 1 mM) d-TC to the organ bath, wait 10 min and repeat the ACh dose–response curve. Finally, add 50 µM (1 mL of 1 mM) d-TC to the organ bath, wait 10 min then repeat ACh dose response. Typical Chart® recordings are shown in Figures 6.1 and 6.2.

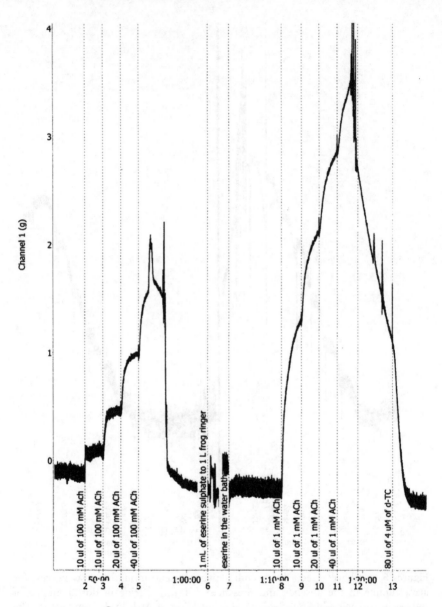

Figure 6.2 Chart® record of antagonism of acetylcholine by d-tubocurarine in the presence of 1 μM serine in leech isolated dorsal muscle. The bath volume was 25 mL, and the abscissa is time (min.).

Results

1. Measure the responses to ACh. Remember to calculate *cumulative* concentration and *cumulative* response.
2. Draw log concentration–response curves (Figure 6.3).
3. Calculate the concentration–shift ratios for ACh alone and ACh with eserine and calculate the extent of potentiation by eserine.
4. Construct a Schild plot for the antagonism of ACh by d-TC and calculate the pA_2 of d-TC acting at the cholinoceptors found in this preparation (Figure 6.4).

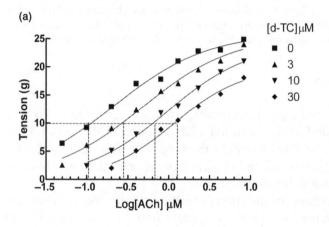

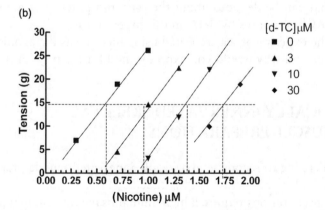

Figure 6.3 Concentration–response curves for acetylcholine (a), and nicotine (b), in the absence and presence of three concentrations of d-tubocurarine (d-TC) in the leech isolated dorsal muscle.

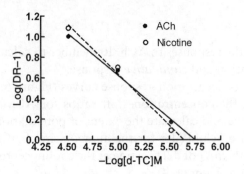

Figure 6.4   Schild plot for the antagonism of responses to acetylcholine and nicotine by d-tubocurarine (d-TC) in the leech isolated dorsal muscle. Note that very similar pA_2 values ($pA_2 = 5.75$, slope $= -1.2$ against acetylcholine, and $pA_2 = 5.64$, slope $= -1.01$ against nicotine) are found for the antagonist using both agonists, indicating that acetylcholine and nicotine are both acting at the same receptor.

Questions

1. Considering these results, what type of receptor do you think is found at the neuromuscular junction of the dorsal muscle of the leech? What would be the expected responses to (a) methacholine, (b) carbachol and (d) depolarizing neuromuscular blockers such as suxamethonium?
2. Account for the effects of eserine on this tissue. What effect would eserine have on the responses found to the drugs listed in Question 1?
3. What can be deduced about the type and potency of antagonism of ACh responses by d-TC in this preparation?
4. If the experiment was repeated using nicotine as the agonist, what value and why would you expect to find for the pA_2 for d-TC?

6.3   FOCALLY INNERVATED SKELETAL MUSCLE PREPARATIONS

It is challenging to recommend a focally innervated muscle preparation that is suitable to be carried out by students in a class practical. Not only do these preparations require a high animal usage (one animal provides no more than two preparations), but also they are technically demanding in that they require good dissection skills and the attachment of a nerve to an electrode. The preparation that is probably most suitable and robust for students is the frog gastrocnemius muscle and sciatic nerve preparation. The nerve is easy to dissect, and the muscle is large and shows a large force development. However, as mentioned in Section 6.2, the access to frogs is restricted since there has been a large decline in frog populations worldwide. Furthermore, it is desirable to minimize the use of vertebrates for student education. Nevertheless, it is important to understand the pharmacology of focally innervated muscles since it is therapeutically important. This is one area where the use of computer aided learning programmes is especially useful. It may be useful to list the types of preparations that have been used in research, and describe the protocol for one in detail. As examples of *in vitro* preparations, the frog gastrocnemius–sciatic nerve, the rat phrenic nerve hemidiaphragm and the innervated biventer cervicis preparation of the chick were particularly widely used. Many years ago, *in vivo* experiments in anaesthetized animals were used to gain much of our understanding about the action of intravenously administered drugs on electrically stimulated muscle preparations. The results of such experiments provided the pharmacological basis for many of the drugs routinely used therapeutically today. Examples of such uses are in surgical operations done under anaesthesia, orthopaedic treatments such as knee and hip replacements, and to control convulsions due to electroconvulsive therapy or infection with *Clostridium tetani* (tetanus).

Compared with multiply-innervated muscles, this type of muscle responds very differently to agonists acting at the nicotinic cholinoceptor and depolarizing and non-depolarizing blockers. In order to observe the effects of drugs on these preparations it is necessary to electrically stimulate the nerve supplying the muscle. Because of the difference in the way tension development is controlled, it is necessary to stimulate using a square-wave pulse generator (Section 3.7) in order to study the pharmacology of focally innervated skeletal muscle. Unlike multiply-innervated muscles, there is only one junction of a motor nerve ending on each muscle cell (neuromuscular junction). Here there is a specialized region where there is a very tight junction between the nerve ending and muscle membrane where there are deep invaginations in the outer muscle cell membrane. The nicotinic cholinoceptors are located in this region. When an action potential arrives at the neuromuscular junction there is release of acetylcholine, it activates post-synaptic nicotinic receptors which result in a small depolarization of the muscle cell, termed miniature end-plate potentials (MEPPs). These are too small to be propagated throughout the fibril. However, if action potentials arrive at the nerve terminal with sufficient frequency, there is an additive effect of the MEPPs and the local end-plate potential depolarization is sufficient to reach the threshold depolarization at about -65 mV. At this voltage, an action potential is generated which is then propagated to the muscle extremities. The resultant wave of depolarization travels to the T-tubule system, where Ca^{2+} is released from sarcoplasmic reticulum storage sites resulting in actin–myosin interaction and the transitory development of tension, termed a twitch. A muscle twitch is defined as a single contraction of a muscle tissue resulting from a single stimulus or a single compound neural action potential. The contraction of a single fibre will give an all-or-nothing transitory twitch, which rapidly returns to the baseline. The amplitude of the twitch increases as the frequency of pulses or pulses increases. This is due to the recruitment or triggering of an increasing number of muscle fibres. At higher frequencies twitches will merge and fail to return to baseline between twitches, and twitches begin to merge. Eventually a frequency is reached where merged twitches reach a maximum. This is termed as tetanus. After a tetanus is established and allowed to return to baseline for 1 min, when stimulation is started at a lower frequency, a supra-normal size of twitch is obtained, termed post-tetanic potentiation. This is thought to be due to a large increase in intracellular $[Ca^{2+}]$ during tetanus, which does not reach baseline during a short rest period. Post-tetanic potentiation rapidly fades over a 10 min period.

To set up a focally innervated muscle preparation, it is first necessary to establish the stimulation parameters. When a sub-maximal voltage and frequency have been established, the actions of drugs acting at the neuromuscular junction can be explored. A large number of drugs act pre-junctionally on the cholinergic nerve endings. This includes drugs interfering with the synthesis, storage and release of acetylcholine. In addition, the pre-junctional nerve is the target of surprisingly large array of venoms and toxins (Bowman and Rand, 1980). The most important of the drugs which act post-junctionally, at sites on the nicotinic cholinoceptor, are depolarizing and non-depolarizing antagonists. Non-depolarizing antagonists compete reversibly for the acetylcholine binding site on the nicotinic receptor. As in multiply-innervated muscle preparations, the block is removed by the application of acetylcholine inhibitors, which cause an increase in the synaptic concentration of acetylcholine. The effect of depolarizing antagonists differs from those on multiply-innervated muscles (Section 6.1). Whereas succinylcholine (suxamethonium) causes the development of a slow contracture in multiply-innervated muscles, in focally innervated muscles there is a short-lasting increase in electrically generated twitches, known as 'fasciculation', which is followed by a block. This block is not reversed by acetylcholinesterase inhibitors, or by any other known drug. This is one disadvantage to the clinical use of depolarizing blockers.

6.3.1 The Frog Gastrocnemius Muscle–Sciatic Nerve Preparation

Experimental Conditions

Organ bath	20 mL, or larger
Ringer solution	Frog Ringer
Aeration	Air
Temperature	20°C
Transducer	Isometric
Dose cycle	10–15 min
Stimulus parameters	Amplitude 0.2–10 V
	Pulse width 0.2 ms
Frequency	0.1–15 Hz

Procedure

Remove the skin from the whole length of the leg. Place the leg in a deep petri dish containing frog Ringer solution. Carefully tease apart the muscle and locate the thick, myelinated sciatic nerve. This is silvery white in appearance. Gently tease out the nerve but be careful not to use any sharp instruments to make sure it stays intact. It is useful to place a narrow glass rod under the rod. Remove the femur. Locate the large gastrocnemius muscle in the lower part of the leg. Note the attachments to the tibia below the knee and the Achilles tendon. Free the muscle from surrounding tissue and tibia, but leave the attachments at the knee and Achilles tendon. A cotton thread is placed around the attachment to the Achilles tendon. This is tied to a metal support. Another thread is placed around the remaining knee tissue and upper muscle attachment and kept free for later attachment to the transducer. The sciatic nerve is carefully fastened to the metal electrode. The whole nerve and muscle structure is placed in the organ bath containing frog Ringer solution, and the free thread from the upper muscle–knee attached to the transducer.

Electrical Stimulation

The maximum amplitude (voltage) and frequency are first established. Begin at a frequency of 0.1 Hz and a voltage of 0.2 V. Stimulate for 3 s intervals, and double the voltage for each interval until a maximum size of twitch is reached. Using this maximum frequency, repeat this procedure of stimulation for 3 s intervals, but this time double the frequency of stimulation. As the frequency increases the twitches will start to fail to return to baseline, and eventually completely merge and reach a maximum size. This is when tetanus has been reached. Rest the tissue for 1 min, and begin to stimulate again at a lower frequency. Large twitches will be seen and that soon fade to a normal height. A graph of the log frequency–response should show a sigmoid relationship.

The actions of blockers of the nicotinic cholinergic receptor are now tested. Select a sub-maximal frequency and stimulate for 3 s. Add 2.5 μM suxamethonium to the bath and wait for 5 min. Stimulate for 3 s intervals, then wash out. When the control twitch height is restored, add 1 μM d-TC and wait 5 min before stimulating for 3 s intervals again. Wash out and add 10 μM eserine, and wait 5 min. Add 2.5 μM suxamethonium, and repeat the electrical stimulation for 3 s intervals. There should be no effect on twitch height. Repeat this procedure with 1 μM d-TC in the presence of eserine. The twitches should be seen to be clearly blocked. After washing out, if 10 μM acetylcholine is added in the presence of serine, the twitches should also be seen to be clearly decreased due to depolarization block.

REFERENCES

Dale, H.H. and Feldberg, W. (1934) Chemical transmission of secretory impulses to the sweat glands of the cat. *J. Physiol.* 82: 121–127.

Flacke, W. and Yeoh, T.S. (1968a) The action of some cholinergic agonists and anticholinesterase agents on the dorsal muscle of the leech. *Br. J. Pharmacol. Chemother.* 33: 145–153.

Flacke, W. and Yeoh, T.S. (1968b) Differentiation of acetylcholine and succinylcholine receptors in leech muscle. *Br. J. Pharmacol. Chemother.* 33: 154–161.

Battacharya, B.K. and Feldberg, W. (1958) Comparison of the effects of eserine and neostigmine on the leech muscle preparation. *Br. J. Pharmacol. Chemother.* 13: 151–155.

Bowman, W.C. and Rand, M.J. (1980) *Textbook of Pharmacology*, Section 17.30, 2nd edn. Oxford: Blackwell Scientific Publications.

7

Isolated Cells

7.1 FRESHLY ISOLATED AND CULTURED CELLS

7.1.1 Advantages of Isolated Cells

The use of suspensions of isolated cells has major advantages over isolated tissues. Numerous assays can be carried out in microtiter plates so that the data for replicate measurements and treatments can be obtained simultaneously. Both time courses and concentration–response curves can be obtained over time periods that are not possible with isolated tissues. This overcomes a drawback with isolated tissues in which responses change with time as tissues age. Microtiter-plate technology requires the availability of specialized microtiter-plate readers and automatic multi-channel pipettes. The large amounts of data generated with these techniques are readily processed with computer software. A limitation in the use of isolated cell techniques is the access to suitable cells. They can be freshly prepared from tissues or may be derived from cell cultures. The least intrusive and most accessible source is different types of blood cells. This does not require the use of animals as blood from human volunteers can be used, provided the health and safety regulations are observed and the necessary ethical approval is given. Isolated cells can be prepared from most organs and tissues, but the problem of heterogeneity of cell types present in whole organs has to be overcome. This requires the use of more complex purification steps. For example, methods for the preparation of longitudinal smooth muscle cells, aortic endothelial cells and parenchymal hepatic cells are available. Fragments of neurones

Practical Pharmacology for the Pharmaceutical Sciences, First Edition. D. Michael Salmon.
© 2014 John Wiley & Sons, Ltd. Published 2014 by John Wiley & Sons, Ltd.

termed synaptosomes can be prepared as a model of synaptic function, as well as astrocyte cell lines.

7.1.2   Cultured Cells

The use of cultured cells presents a number of advantages, but there are also limitations. They are convenient and have a theoretically unlimited availability. Furthermore, they do not require any special licences for the use of animals. A cultured cell line is often chosen as a model for a particular type of cell found *in vivo*. For example, embryonic stem cell derived megakaryocytes have been cultured as a model for platelets, and differentiated HL-60 cells are used as a model for neutrophils. Their functions resemble the native cells to varying extents. An alternative is to use primary cell cultures, which are of limited value, as their function depends on the age and history of the donors, and they do not behave metabolically as cells *in vivo*. Primary cultured cells multiply slowly and age rapidly. Transformed cell lines are virtually immortal and are well characterized, so that experiments may be carried out over years, and results from different laboratories can be compared. However, they are expensive to maintain, both in time and resources, and must continually be validated to ensure their morphology and function. Cell culture requires a great deal of skill to develop sterile techniques. It is also very time consuming, and so may not be recommended for relatively short student pharmacology projects if ample technical help is not available.

The most suitable cell types that are suitable for class experiments or short projects are blood-derived cells, as animal licences are not required, and can be handled without specialized sterility techniques. Examples of two such cell types are platelets and neutrophils. They can both be prepared rapidly in large numbers and are easy to manipulate. Experiments carried out with these two cell types are discussed here.

7.1.3   Cell Counting

Isolated cell suspensions can not only be counted to determine the cell density of the suspension or the number in each well of a microtiter dish, but also information about the cell size distribution and their characteristics can be measured. The simplest method of cell counting is the use of a haemocytometer under a microscope. The number of live cells can be counted by identifying the cells that exclude the dye trypan blue. Other more sophisticated methods are the use of Coulter counters

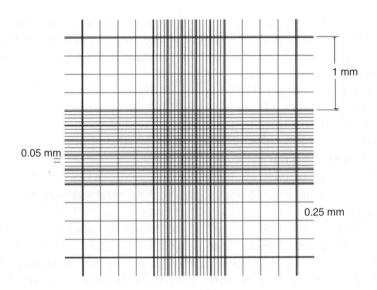

Figure 7.1   Calibrated Neuberger haemocytometer.

and flow cytometry. Other methods used to quantify bacterial growth, such as counting colonies or measuring optical density, are not generally applicable to mammalian cells or cell lines.

Use of the Haemocytometer

A haemocytometer is a specialized slide upon which has a finely calibrated grating etched into its surface. When a cover slip is placed over the grating, there are a large number of cubes of known volume. The improved Neuberger haemocytometer has a grating as illustrated in Figure 7.1. The grating is divided into nine large squares, each measuring 1×1 cm. Inside each large square are nine smaller squares measuring 0.25×0.25 cm. The centre large square is divided into 25 smaller squares each measuring 0.2×0.2 mm. Each of these squares contains 16 min squares measuring 0.05×0.05 mm. The depth of the chambers when a tightly fitting cover slip is placed is written on the haemocytometer, and is usually 0.1 mm.

To count the number of live cells in a suspension, the following procedure is followed:

- The haemocytometer is cleaned with alcohol and dried using tissue paper and lens tissue.

- The surface of the haemocytometer is moistened with a slightly damp tissue and a cover slip is placed over the haemocytometer. A tight fit has been achieved when coloured refraction rings can be seen.
- Equal volumes of a 0.4% trypan blue dye and the cell suspension are mixed together. Only a very small volume is needed, for example, 50 μL of each in an Eppendorf plastic centrifuge tube.
- 10 μL of this suspension is placed at the edge of the cover slip and allowed to fill the wells.
- Under a phase contrast inverted microscope, any blue cells are dead; whilst the live cells do not take up the dye appear as bright, refractile cells.
- A large number of cells (100–150) should be counted to ensure a reliable figure. This is done using a tally clicker. The number of large squares (1 × 1 cm) that need to be counted depends on the cell density. If there are more than 100–150 cells in the four large corner squares, the cell suspension should be diluted and recounted. Platelets should be counted at 40 × magnification. Even at this magnification, platelets will appear very small. They have very heterogeneous size distribution, but they have a diameter of less than 3 μm.
- Calculation of the cell concentration is performed as follows. The volume of one large square is $1 \times 1 \times 0.1 = 0.1$ mm^3. Divide the total number of cells counted by the number of large squares counted. The number of cells per large square is multiplied by 10^4 to find the number of cell/mL. This figure must be multiplied by two to account for the dilution with trypan blue. If the cell suspension was diluted before placing in the haemocytometer, this must also be taken into account.

Coulter Counters and Flow Cytometry

This is the most rapid and informative method of performing a cell count. It is based on the principle that when a particle, such as a cell suspended in an electrolyte solution, is drawn through a narrow aperture across which there is an electric field, there is an increase in resistance which is proportional to its volume. The consequent transitory fall in current flow as cells pass through the aperture enables a record of cell number and cell volume to be made. There are two formats used in Coulter counters, aperture and flow formats. In the aperture format, electrodes are placed on each side of an aperture, the size of which is fixed according to the size of the cells to be counted such that only single cells can pass through the aperture. Cells suspended in a physiological buffer flow through the aperture by applying vacuum on one side of the aperture. Whilst this method is relatively simple, it suffers from the disadvantage that it is a batch method and only one suspension can be measured at a time. This is the most suitable method for the experiments described in this section. The flow format is a continuous method, where electrodes are placed at both ends of a tube. As cells flow through the tube they cause a fall in resistance and current flow, which is monitored and the data processed using appropriate software.

Flow cytometry was developed from flow format Coulter counters. A focussed stream of cells crossed a laser beam of light of fixed wavelength. Light beams scattered in two directions are detected. Forward scattered light travels in the same direction as the laser beam. An obscuration baffle is placed in the path of the laser beam to prevent it from striking the photomultiplier tube (PMT), and only the forward-scattered light passing round the edges of the baffle are detected. If a fluorescent probe is attached or inserted into the cells, light emitted at 90° to the direction of the light beam, is detected by additional PMTs. Filters placed in front of the PMTs ensure that light of only one wavelength is detected by each PMT. This technique is by no means limited to determining the number and size distribution of cells. Flow cytometry is now an exceptionally powerful technique throughout the life sciences, as it can measure an increasing amount of information about events in the extra- and intra-cellular environment. Flow cytometry can also be used as a preparative technique to separate different cells of different types or characteristics.

Pharmacological responses can also be monitored in isolated cell suspensions using Coulter counter and flow cytometry techniques. Contraction of a suspension of isolated intestinal smooth muscle cells results in a shape change that has been monitored using a Coulter counter (Singer and Fay, 1977). This system is particularly suited to the elucidation of signal transduction events over very short times preceding contraction (Salmon and Honeyman, 1980).

7.2 PLATELETS

Platelets are the second most numerous cell type found in blood. They have a critical role in haemostasis. Physiologically, platelets are activated in response to a damaged blood vessel. They aggregate in attempt to seal the breach, as well as release numerous compounds that cause vasoconstriction and accelerate aggregation by a positive feedback mechanism.

Aggregation of platelets is initiated by exposure to sub-endothelial collagen, thrombin or eicosanoids (principally thromboxane A_2), platelet activating factor (PAF), 5-hydroxytryptamine and purine nucleotides, principally ADP. Pathologically, platelet activation occurs as is an early step in the formation of an arterial thrombus. Therefore anti-platelet drugs, such as aspirin, are invaluable as prophylactic treatments to prevent arterial thrombi.

Platelet aggregation is classically measured by monitoring the fall in the optical density of platelet suspensions in an aggregometer. This may be done either by using platelet-rich plasma (which will contain plasma proteins including coagulation factors), or using washed platelet suspensions in a physiological buffer. In this experiment washed platelets suspended in HEPES-buffered phosphate buffer (pH 7.4) were used. In an aggregometer, a cloudy platelet suspension is stirred in a cuvette and maintained at 37°C and placed in the light path of a colorimeter. The machine measures the fall in optical density as the cloudy platelet suspension forms large clumps or aggregates, thus allowing more light to pass through a cuvette. This is a time-consuming procedure, as only one cuvette is measured at a time, although some machines allow up to six cuvettes to be studied simultaneously. A major defect of this technique is that platelets deteriorate and change their characteristics over several hours of the experiment.

An alternative and more rapid technique is to use a microtiter-plate reader (Salmon, 1996). This is more suited to class experiments as it obviates the need to invest in large numbers of aggregometers, and the microtiter-plate reader is a multi-functional machine. These basically consist of a plastic dish with 96 wells, each one of which acts as a cuvette – so the optical density of 96 samples can be read at the same time. Using a kinetic microtiter-plate reader, not only can this be done, but also the temperature of the samples can be maintained at 37°C and the samples can be agitated to keep the samples suspended and ensure platelet–platelet contact. Thus it is an important improvement over traditional aggregometers. Using the microtiter-plate reader, large amounts of data are rapidly generated, necessitating the use of computers for analysis.

Platelet Preparation

Blood is taken from healthy volunteers who must be refrained from taking any non-steroidal anti-inflammatory drugs such as aspirin during the 2 weeks before venous blood was taken. The blood is transferred immediately into four tubes containing 3 mL of acid citrate dextrose anticoagulant (trisodium citrate dihydrate 25 g/L, citric acid monohydrate 15 g/L, glucose 20 g/L). Centrifuge of the blood was at 800 g for 5 min, which results in sedimentation of the red blood cells and a supernatant, which was cloudy straw-coloured platelet rich plasma (PRP). The PRP is removed carefully without disturbing the red blood cells. The PRP is then centrifuged at 800 g for a further 20 min, after this the clear platelet poor plasma is discarded from the top, and the platelets are carefully resuspended in HEPES saline buffer (pH 7.4) containing the following (mM): NaCl 145, KCl 5, MgCl$_2$ 1, 6H$_2$O 1, HEPES (N-2-hydroxyethylpiperazine-N-2-ethansulphonic acid) 10, and D-glucose 10. The washed human platelets are stored at room temperature, and remain viable for several hours.

Measurement of Platelet Aggregation

To measure platelet aggregation, 150 μL of washed platelet suspension is dispensed into the 96-well microtiter plates (flat-bottomed 0.3 mL capacity). The microtiter-plate reader should be capable of being programmed to make kinetic measurements, and be temperature controlled and have shaking or stirring facilities. Examples of these are the Labsystems iEMS reader/dispenser MF and the Multiskan (from Thermo Scientific), interfaced with a computer controlled by Genesis or SkanIt software. The machine is programmed to incubate the plate for 5 min at 37°C with or without inhibitor, before 8 μL of an agonist such as collagen, arachidonic acid or thrombin was dispensed into the wells. Optical density readings at 620 nm, taking less than 10 s for the entire plate are taken, followed by orbital agitation at 1400 rpm for the next 21 s, and then read again. This cycle is repeated for the next 5 min period. The computer is programmed to automatically plot the time course of absorption in all 96 wells in real time, and all readings are stored on the computer's hard drive.

7.2.1 Inhibition of Aggregation by Nitric Oxide Donors

The aim of this experiment was to investigate the inhibition of platelet aggregation by two compounds that are known to generate nitric oxide (NO). The classical text-book view is that inhibition of platelet

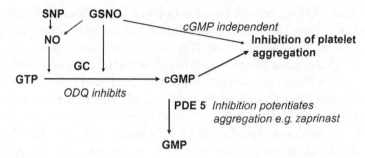

Figure 7.2   Scheme to illustrate how ODQ and zaprinast modulate the inhibition of platelet aggregation by the nitric oxide donors, SNP and GSNO. In addition, extracellular HbO_2 tightly binds to NO preventing its action.

aggregation (like smooth muscle relaxation) is brought about by the activation of guanylate cyclase by NO, and this is essential for NO to exert its effect. However not all experimental data agrees with this conclusion, and it is now clear that there is a cyclic GMP independent mechanism (Crane *et al.*, 2005). The two NO donors used are sodium nitroprusside (SNP) and nitrosoglutathione (GSNO). To investigate the mechanism of these compounds two inhibitors that interfere with the generation of NO are used. ODQ ($1H$-[1,2,4]oxadiazolo[4,3,α]quinoxalin-1-one) inhibits guanylate cyclase, and zaprinast inhibits the breakdown of cyclic GMP by cGMP specific phosphodiesterase (PDE5). In addition, NO binds tightly to oxyhaemoglobin, effectively removing it from solution and preventing its action (Figure 7.2).

Protocol

Layout of Microtiter Plate   Layouts for Plates 1 and 2 are shown in Tables 7.2 and 7.4. For plate 3, the effect of oxyhaemoglobin on the inhibition of arachidonic acid stimulated aggregation by SNP and GSNO. Platelets are treated in triplicate as in plates 1 and 2. Columns 1–6 contain SNP at the same concentrations as indicated for plate 1 (Table 7.1), and columns 6–12 are treated with the same concentrations of GSNO as indicated for plate 2 (Table 7.3). Columns 4–6 and 10–12 contain oxyhaemoglobin (oxyHb) at 10 μM.

Table 7.1   Sodium nitroprusside (SNP) concentrations (nM) for the concentration–response curve in plate 1.

SNP1	SNP2	SNP3	SNP4	SNP5	SNP6	SNP7
1000	500	100	50	10	5	1

Table 7.2 Plate 1. 96-well plate layout for the inhibition of arachidonic acid (AA) stimulated platelet aggregation by nitroprusside (SNP), in the absence or presence of ODQ (10 μM) and/or zaprinast (10 μM). All treatments are carried out in triplicate. After 5 min incubation with inhibitors, AA (6.25 μM) is added to all wells.

	1	2	3	4	5	6	7	8	9	10	11	12
A		AA only			ODQ		Zaprinast (Z)			ODQ + Z		
B		SNP7			SNP7 + ODQ			SNP7 + Z			SNP7 + ODQ + Z	
C		SNP6			SNP6 + ODQ			SNP6 + Z			SNP6 + ODQ + Z	
D		SNP5			SNP5 + ODQ			SNP5 + Z			SNP6 + ODQ + Z	
E		SNP4			SNP4 + ODQ			SNP4 + Z			SNP4 + ODQ + Z	
F		SNP3			SNP3 + ODQ			SNP3 + Z			SNP3 + ODQ + Z	
G		SNP2			SNP2 + ODQ			SNP2 + Z			SNP2 + ODQ + Z	
H		SNP1			SNP1 + ODQ			SNP1 + Z			SNP1 + ODQ + Z	

Table 7.3 The nitrosoglutathione (GSN) concentrations (nM) used in plate 2.

GSNO1	GSNO2	GSNO3	GSNO4	GSNO5	GSNO6	GSNO7
1000	500	100	50	10	5	1

Table 7.4 Plate 2. 96-well plate layout for the inhibition of arachidonic acid (AA) stimulated platelet aggregation by nitrosoglutathione (GSNO), in the absence or presence of ODQ (10 μM) and/or zaprinast (10 μM). After 5 min incubation with inhibitors, AA (6.25 μM) is added to all wells.

	1	2	3	4	5	6	7	8	9	10	11	12
A		AA only			ODQ		Zaprinast (Z)			ODQ + Z		
B		GSNO7			GSNO7 + ODQ			GSNO7 + Z			GSNO7 + ODQ + Z	
C		GSNO6			GSNO6 + ODQ			GSNO6 + Z			GSNO6 + ODQ + Z	
D		GSNO5			GSNO5 + ODQ			GSNO5 + Z			GSNO6 + ODQ + Z	
E		GSNO4			GSNO4 + ODQ			GSNO4 + Z			GSNO4 + ODQ + Z	
F		GSNO3			GSNO3 + ODQ			GSNO3 + Z			GSNO3 + ODQ + Z	
G		GSNO2			GSNO2 + ODQ			GSNO2 + Z			GSNO2 + ODQ + Z	
H		GSNO1			GSNO1 + ODQ			GSNO1 + Z			GSNO1 + ODQ + Z	

Questions

1. Plot log concentration–inhibition curves for plates 1 and 2. Note that all treatments were carried out in triplicate. Plot the mean ± SEM ($n = 3$). Calculate IC_{50} values and tabulate them. An example of typical results is shown in Figure 7.3.

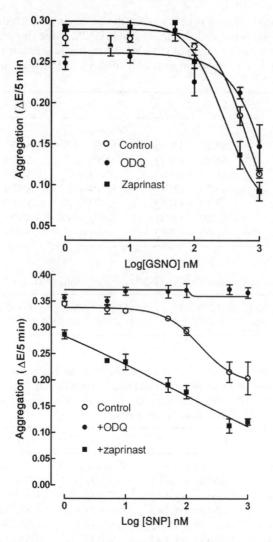

Figure 7.3   Inhibition of AA-stimulated platelet aggregation by SNP and GSNO in the presence and absence of ODQ and zaprinast. Each point indicates the mean of three observations and the error bars indicate the SEM.

2. Draw up a flow chart to indicate the sites of action of ODQ, zaprinast and oxyHb. Give your reasons for thinking that the above data support or contradict the contention that NO donors all act via cyclic GMP.

3. What alternative explanations might there be for the mode of action of NO donors?

7.3 NEUTROPHILS

Neutrophils, or polymorphonuclear neutrophils (PMN), are the most numerous of the leukocyte fraction of human plasma. They are identified histologically with haematoxylin–eosin dye, when the cytoplasm stains pale blue with dark blue multi-lobed nuclei. They are normally present in blood at between 2.5 and 7.5×10^9/L. They are the primary responders to inflammatory signals such as IL-8, leukotriene B_4, PAF, the peptide formylated-met-leu-phe (FMLP, fMLP or fMLF) and the protein kinase C (PK-C) activator phorbol-12-myristate-13-acetate (PMA). Characteristic primary responses are a rise in intracellular $[Ca^{2+}]$ followed by a rapid uptake of O_2 (oxidative or respiratory burst) produced by activation of the NADPH complex and myeloperoxidase that causes the accumulation of toxic reactive oxidative species (ROS), such as superoxide and hypochlorite. This is followed by chemotaxis and phagocytosis of bacteria. The activation of neutrophils is complex and not fully understood. FMLP, PMA and macrophage-derived growth factors (such as GM-CSF and TNFα) can activate superoxide production through activation of NADPH oxidase, other stimuli such as PAF that act weakly on their own have a priming or potentiating effect on primary stimulators, such as FMLP (Kitchen et al., 1996) The relationship between activation of NADPH oxidase and the rapid increase in intracellular $[Ca^{2+}]$ is obscure. FMLP strongly stimulates superoxide generation and the rise in $[Ca^{2+}]$, as do PAF and LTB$_4$, yet PMA does not produce an increase in intracellular $[Ca^{2+}]$, and inhibits the rise produced by FMLP, PAF and LTB$_4$. It appears that whilst $[Ca^{2+}]$ and PK-C activation both stimulate NADPH oxidase, PK-C has a negative feedback effect to inhibit the increase in intracellular $[Ca^{2+}]$.

It has been suggested that drugs that specifically inhibit ROS formation may be useful in treating cardiovascular diseases caused by oxidative stress, such as arteriosclerosis (Stocke and Keaney (2004); Wind et al., 2010).

However, it should be noted that subjects who possess a mutation in a component of NADPH oxidase show symptoms of chronic granulomatous disease (CGD), in which sufferers are very prone to bacterial and fungal infections.

Preparation of Neutrophils

10 mL dextran is added to 50 mL of freshly drawn heparinized blood and cells are allowed to sediment under gravity for 60–90 min. The

supernatant is removed and carefully layered above 10 mL Histopaque 1119-1 (Sigma–Aldrich). After centrifugation at 2000 rpm. for 10 min at room temperature, the supernatant and Histopaque are removed leaving sedimented neutrophils and a small contamination of erythrocytes, which is removed by hypotonic lysis. Ten millilitres of distilled water are added and after 20 s 10 mL of hypertonic (two times concentrated) phosphate-buffered saline (PBS) are added and vigorously agitated to restore the tonicity. Neutrophils are sedimented at 2000 rpm. for 5 min and resuspended in 1 mL PBS containing 5 mM glucose without $CaCl_2$, and counted using a haemocytometer. Neutrophil suspensions are adjusted to a cell density of 10^7/mL.

7.3.1   Measurement of NADPH Cytochrome c Reductase

Several methods for measuring the respiratory, or oxidative, burst in activated neutrophils have been used (Dahlgren and Karlsson, 1999). The direct uptake of O_2 can be measured using a Clarke electrode, but this is time consuming. More frequently, methods to measure superoxide formation have been used. Examples of these are reduction of ferricytochrome c (to measure extracellular superoxide formation), chemiluminescence using luminol, dichlorofluorescein fluorescence (to measure intracellular superoxide formation) and electrochemical detection (Tarpey and Fridovich, 2001). The enzymic method that measures reduction of ferricytochrome c utilizing microtiter-plate technology is described here.

Ferricytochrome c Reduction

In this method, extracellular superoxide production is measured spectrophotometrically by the increase in absorption at 550 nM as Fe^{3+} cytochrome c is reduced to Fe^{2+} cytochrome c:

$$Fe^{3+}cytc + O_2^- \rightarrow Fe^{2+}cytc + O_2$$

The extinction coefficient of ferrocytochrome c at 550 nm is 3.36 times greater than that of ferricytochrome c. However, there are several potential sources of interference with this reaction when it is carried out using neutrophil suspensions. Cytochrome c reduction is not specific for superoxide. The presence of any anti-oxidizing chemicals can also reduce ferricytochrome c. Specificity of this reaction for superoxide is achieved in part by the extent of inhibition of cytochrome c reduction by added superoxide dismutase (SOD). The reaction can be carried out using a double-beam spectrophotometer, where the reference cell contains the reaction mixture containing SOD. Alternatively, the reaction can be carried out in a 96-well microtiter dish, where reactions are carried out with and without SOD.

Protocol

The experiment demonstrates the activation of NADPH oxidase by three agonists, FMLP, Leukotriene B_4 and PAF. The enzyme can be shown to be inhibited by diphenylene iodonium (DPI) and apocynin (Wind et al., 2010).

The kinetic assay is carried out in 96-well microtiter plates. Each well has a final volume of 200 μL and each treatment includes a well without cytochrome c and another with cytochrome c. The superoxide production is calculated from the difference in $\Delta E_{550\ nm}$ in the two cells. Wells contain:

- 20 μL of a neutrophil suspension (2×10^7 cells/mL), final well concentration 2×10^6 cells/mL;
- 100 μL of 100 μM ferricytochrome c (horse heart, M.W. 12,327), final well concentration 50 μM;
- 20 μL of 1 mg/mL SOD solution, when required, well concentration 100 μg/mL;
- 20 μL agonist and/or inhibitor (stock concentration 20 × final);
- Well volume is made up to 200 μL with phosphate buffered saline (PBS) containing 10 mM glucose.

The extinction at 550 nm is sampled at 20 s intervals sample over time (5 min for FMLP, 10 min for PMA), and ΔE/min calculated from the linear portion of the time course. The rate of superoxide production is defined as the SOD-inhibitable activity, and is obtained by subtracting the ΔE/min with SOD from ΔE/min without SOD. Superoxide production is calculated from the extinction by using an extinction coefficient of reduced cytochrome c of 2.11×10^4 M/cm, and knowing the light path length in the microtiter well. From the Beer–Lambert law:

$$\text{Extinction} = \varepsilon CL,$$

where ε is the molar extinction coefficient (M^{-1} cm^{-1}), C is the concentration (M), and L is the light path length (cm).

7.3.2   Measurement of Intracellular [Ca^{2+}]

The control of intracellular [Ca^{2+}] is a universal signal transduction mechanism found throughout biology. Calcium regulates a wide range of vital cell functions including enzyme activities, attachment, motility, morphology, metabolic processes, cell-cycle progression, signal transduction, replication, gene expression and electrochemical responses. Intracellular [Ca^{2+}] is regulated by two basic processes: mobilization of Ca^{2+} from intracellular stores in the endoplasmic reticulum, and influx and efflux through the cell membrane. Receptor activation results in a rapid primary transient increase or spike in intracellular [Ca^{2+}], followed by a later maintained influx of Ca^{2+} through cell surface ion channels, which may be either voltage or receptor-operated. The initial transient spike in intracellular [Ca^{2+}] is triggered by activation of receptors linked to a phospholipase C isoenzyme that breaks down a trace membrane phospholipid, phosphatidylinositol 4,5-bisphosphate to release the water-soluble second messenger inositol 1,4,5 trisphosphate (InsP$_3$ or IP$_3$) and lipid-soluble diacylglycerol (DAG). IP$_3$ binds to specific receptors on the endoplasmic membrane to release stored Ca^{2+}. The secondary, maintained influx of Ca^{2+} is brought about as a consequence of store depletion as described by the capacitance model of Putney (1986, 2010), and, in some cell types at least, by the activation of receptor-operated Ca^{2+} channels. Intracellular [Ca^{2+}] can be measured by several techniques. Cells can be loaded with an intracellular fluorescent or luminescent probe which is sensitive to [Ca^{2+}]. There is a large variety of such probes for different applications, such as are Aequorin (a luminescent phosphoprotein from a jellyfish), quin2, fluo-, fura- and indo-based compounds, which are derived from the structure of the calcium-specific chelator EGTA. The fluorescent signal can be detected by conventional fluorimetry or more sophisticated methods such as confocal microscopy or flow cytometry.

In the method described here, it is possible to measure separately the two phases of store release and influx. Neutrophils are loaded with the calcium probe, Fura2. The release of stores is measured as a spike in fluorescence when an agonist is added in the absence of extracellular Ca^{2+}. CaCl$_2$ is then added and a maintained increase in fluorescence is seen reflecting the Ca^{2+} influx.

Measurement of Intracellular [Ca²⁺] in Neutrophils Using Fura2

Measurement of Intracellular $[Ca^{2+}]$ in Neutrophils Using Fura2

Intracellular $[Ca^{2+}]$ is measured by monitoring the fluorescence of Fura2 by an adaption of the ratiometric method introduced by Grynkiewicz *et al.* (1985). Here the neutrophils (or other isolated cells, such as platelets) are loaded with the fluorescent calcium chelator, Fura2 by incubating the cells with the cell-permeant acetoxymethyl ester of Fura2, Fura2 AM. This ester is cleaved intracellularly by esterases leaving the free acid trapped inside cells. Fura2 has a different excitation maximum for the Ca^{2+}-free acid (380 nm) and the Ca^{2+}-chelated compound (340 nm), but both have the same emission maxima (509 nm). The most sensitive measure of $[Ca^{2+}]$ is obtained by measuring the ratio of fluorescence at excitation wavelengths of 340 nm and 380 nm (Figure 7.4).

This has a major advantage over earlier calcium probes, such as Quin2 that measures fluorescence at a single wavelength, because if the cells settle out during measurement there is an artifactual change in signal. Fura2 does not suffer from this because there is no change in ratio as cells settle.

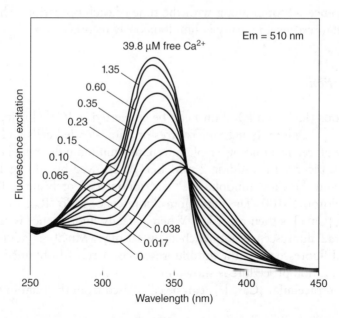

Figure 7.4 Fluorescence excitation spectra of Fura2 measured at an emission wavelength of 510 nm.

Protocol

Neutrophils as prepared above are sedimented at 2000 rpm. for 5 min and resuspended in 4 mL PBS containing 5 mM glucose and 1 mM $CaCl_2$. Eight microlitres of 1 mM Fura2 AM in DMSO (final concentration 2 µM) are added, and tubes were covered in aluminium foil and incubated for 30 min at 37°C in a water bath. Neutrophils are then sedimented at 800 g for 5 min, and resuspended in PBS containing 5 mM glucose to provide a cell density of about 1.8×10^7/mL.

It is more rapid and economical to use a programmable microtiter-plate fluorescence reader (e.g. Fluoroskan Ascent FL, Thermo Scientific), rather than a double-beam spectrofluorimeter, the Ascent microtiter-plate fluorescence reader. The reader should be capable of operating at 37°C and have a facility for agitating the cells in the micrititer-plate after each addition. The sequence of operations is varied according to the requirements of each experiment, and this is achieved by programming using the Ascent software. Experiments typically consist of three measurement stages (basal, Ca^{2+} release by agonist and Ca^{2+} influx after addition of $CaCl_2$). Measurement consists of recording fluorescence intensity at two excitation wavelengths (340 nm and 380 nm) and a single emission wavelength (510 nm). In the record of the raw data, the fluorescence is stored along with the time of each operation. The ratio of the fluorescence intensity is simultaneously recorded.

Calibration

The ratios (F340 nm/F380 nm) can be converted to $[Ca^{2+}]$ using a calibration procedure. Ninety microlitres of Fura2-loaded cells and 10 µL PBS are placed in a microtiter-plate well, the minimum fluorescence ratio (R_{min}) is recorded by adding 2 mM EGTA followed by adding 10 mM $CaCl_2$ and the Ca^{2+}-ionophore, 1 µM ionomycin, or lysing cells with 0.1% Triton X-100. The maximum fluorescence ratio (R_{max}) is at saturating $[Ca^{2+}]$ is then recorded. When a steady fluorescence is reached, the Fura2 fluorescence is quenched by 10 mM, $MnCl_2$ is added. The residual fluorescence is the autofluorescence. A record obtained during the calibration procedure is shown in Figure 7.5.

The intracellular $[Ca^{2+}]_{free}$ can be calculated from the relationship:

$$\left[Ca^{2+}\right] = K_d \times \frac{(R - R_{min})}{(R - R_{max})} \times \frac{S_{f2}}{S_{b2}}$$

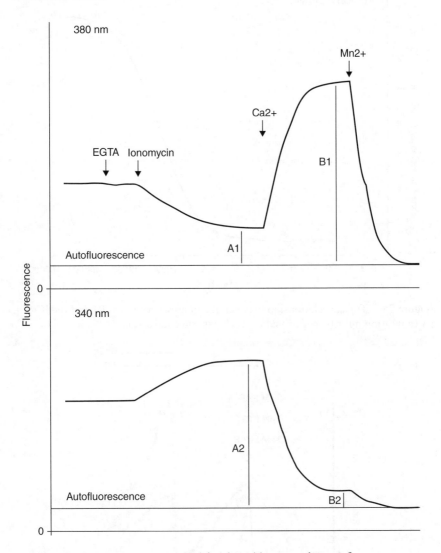

Figure 7.5 Parameters measured for the calibration of Fura2 fluorescence.

where

$$R_{min} = \frac{A1}{A2}; \ R_{max} = \frac{B1}{B2} \ and \ \frac{S_{f2}}{S_{b2}} = \frac{A2}{B2}$$

and K_d is the dissociation constant of Ca^{2+} for Fura2, and is taken as 225 at 37°C. A graphical representation of the relationship between the ratio of F380 nm/R340 nm is shown in Figure 7.6.

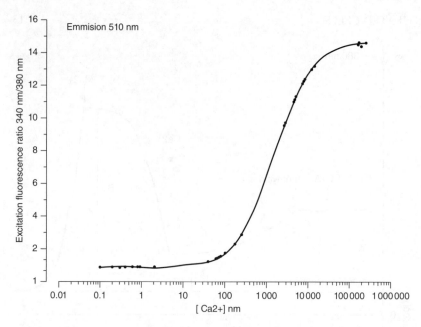

Figure 7.6 Typical relationship between the fluorescence ratio (R) and the $[Ca^{2+}]$. A calibration must be performed for each experimental setup.

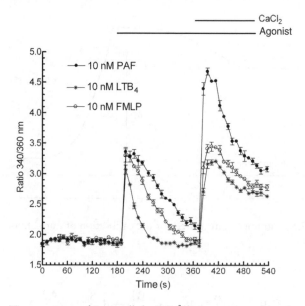

Figure 7.7 Time courses of intracellular $[Ca^{2+}]$ (as measured by the ratio of the excitation fluorescence at 380 nm and 340 nm, and emission at 510 nm) in Fura2-loaded neutrophils stimulated with FMLP, LTB$_4$ and PAF. Each point is the mean of six observations, and the error bar indicated the SEM. The first peak shows the calcium transient caused by the release of bound calcium from stores in the sarcoplasmic reticulum, and the second rise indicates the influx of Ca^{2+} after the addition of extracellular $CaCl_2$.

Ninety microlitres of neutrophil suspension (2×10^7/mL) and 5 μL of 20 mM EGTA are added to each well of a flat-bottomed, black 96-well microtiter plate. The cells are allowed to warm up to 37°C for 5 min, and then the baseline fluorescence is recorded. This is followed by 10 μL of agonist and/or inhibitor and orbital shaking for 5 s at 1200 rpm. Each reading is integrated for 20 ms and a plate acceleration of 10 ms is used. An increase in the ratio is seen followed by decay back to baseline levels. This shows the release of calcium from stores. When the baseline is re-established, 2 mM $CaCl_2$ is added and readings continued until the ratio reaches a plateau is reached. This shows the extent of calcium influx. A record of the time courses of the increase in intracellular $[Ca^{2+}]$ obtained for three agonists, FMLP, PAF and LTB_4 are shown in Figure 7.7.

REFERENCES

Crane, M.S., Rossi, A. G., and Megson, I.L. (2005) A potential role for extracellular nitric oxide generation in CGMP-independent inhibition of human platelet aggregation: biochemical and pharmacological considerations. *Bri. J. Pharmacol.* 144: 8498.

Dahlgren, C. and Karlsson, A. (1999) Respiratory burst in human neutrophils. *J. Immunol. Methods.* 232: 3–14.

Grynkiewicz, G., Poenie, M., and Tsien, R.Y. (1985) A new generation of Ca^{2+} indicators with greatly improved fluorescence properties. *J. Biol. Chem.* 260: 3440–3450.

Kitchen, E., Rossi, A.G., Condliffe, A.M., Haslett, C., and Chilvers, E.R. (1996) Demonstration of reversible priming of human neutrophils using platelet-activating factor. *Blood* 88: 4330–4337.

Putney, J.W. (1986) A model for receptor-regulated calcium entry. *Cell Calcium* 7: 1–12.

Putney, J.W. (2010) Pharmacology of store-operated calcium channels. *Mol. Interventions.* 10(4): 209–218.

Salmon, D.M. (1996) Optimization of platelet aggregometry utilizing microtiterplate technology and integrated software. *Thromb. Res.* 84: 213–216.

Salmon, D.M. and Honeyman, T.W. (1980) Proposed mechanism of cholinergic action in smooth muscle. *Nature* 284: 344–345.

Singer, J.J. and Fay, F.S. (1977) Detection of contraction of isolated smooth muscle cells in suspension. *Am. J. Physiol.* 232: C138–C143.

Stocke, R., and Keaney, J.F., Jr (2004) Role of oxidative modifications in atherosclerosis. *Physiol. Rev.* 84: 1381–1478.

Tarpey, M.M. and Fridovich, I. (2001) Methods of detection of vascular reactive species: nitric oxide, superoxide, hydrogen peroxide, and peroxynitrite. *Circ. Res.* 89: 224–236.

Wind, S., Beuerlein, K., Eucker, T., Müller1, H., P Scheurer, P., Armitage, M.E., Ho, H., Schmidt, H.H.H.W. and K Wingler, K.(2010) Comparative pharmacology of chemically distinct NADPH oxidase inhibitors. *Br. J. Pharmacol.* 161: 885–898.

8

Biochemical Pharmacology

8.1 PHARMACOLOGICAL APPLICATIONS OF COMMON BIOCHEMICAL TECHNIQUES

In the search for a deeper understanding of the mechanism of action of drugs, pharmacology has recruited techniques from all branches of the biosciences. In attempts to find the link between receptor occupation and the production of a cellular response, the focus of research on the action of neurotransmitters and other modulators has moved from plasma membrane receptors to the interior of the cell. The void previously labelled 'efficacy' is now filled by increasing understanding of signal transduction mechanisms. A wide range of biochemical techniques are used to study the biochemical events after receptor activation. Biochemical techniques are needed to study intracellular signal transduction processes, such as second messenger concentrations, activation of protein phosphorylation cascades and nuclear gene transcription. The number of techniques that are suitable for large class practicals are somewhat constrained due to safety or costs. Assays for second messengers, such as cyclic nucleotides and inositol trisphosphate, that were initially based on the use of radioisotopically labelled ligands have largely been replaced by immunological methods in which fluorescence or bioluminescence is detected. The use of radioactively labelled compounds has been discontinued in many teaching laboratories because of the safety concerns and the high costs of disposal of radioactive waste. The alternative methods utilize expensive reagents, such as monoclonal antibodies for ELISA based methods. The prohibitively high cost prevents many teaching institutions from including them in undergraduate

Practical Pharmacology for the Pharmaceutical Sciences, First Edition. D. Michael Salmon.
© 2014 John Wiley & Sons, Ltd. Published 2014 by John Wiley & Sons, Ltd.

practical courses, but can be included in smaller classes for post-graduate courses. Most assays utilizing antibodies and fluorescent or bioluminescent tags to determine concentrations are available as commercial kits. Some techniques can be easily included in practical courses, such as chromatography, electrophoresis and enzymology. Simple chromatography techniques, such as high-performance thin layer chromatography (HPTLC), are eminently suitable for demonstrating the metabolism of drugs. Other techniques such as HPLC and electrophoresis are not readily available for teaching classes of more than ten students, since they require expensive equipment or detection reagents. However, enzymological techniques stand out above all others in their suitability for inclusion in practical courses for large numbers of students. Enzymes are an important target of many drugs, indeed they are sometimes referred to as the 'receptors' for some classes of drugs, although this can be a rather confusing terminology.

8.2   ENZYME INHIBITORS

There is a long list of widely prescribed drugs that act as enzyme inhibitors. Examples are non-steroidal anti-inflammatory drugs and angiotensin-converting enzyme inhibitors. Here the assay of four enzymes which are the target of important classes of drugs is described, acetylcholinesterase (AChE), monoamine oxidase (MAO), Na^+,K^+-ATPase and thrombin. AChE inhibitors are therapeutically important because they are indirectly acting parasympathomimetics. Neostigmine is used to reverse the action of some non-depolarizing nicotinic receptor blockers (muscle relaxants) that are used during anaesthesia. It can also improve the muscle tone of patients with myasthenia gravis. Centrally acting anticholinesterase inhibitors (such as donepezil) are used to control Alzheimer's disease. MAO inhibitors were an early therapy for depression, and are also used in the treatment of Parkinson's disease. Cardiac glycosides classically inhibit Na^+,K^+-ATPase and were previously used in the treatment of heart failure, although now they are primarily used to treat supraventricular arrhythmias. Thrombin is a protease that has a central role in haemostasis. Directly acting thrombin inhibitors have been developed to prevent coagulation in certain thromboembolic conditions, chiefly venous thromboembolism and heparin-induced thrombocytopenia.

8.3 ACETYLCHOLINESTERASE INHIBITORS

AChE rapidly catalyses the inactivation of acetylcholine released from cholinergic nerve terminals. It is inhibited by a number of competitive, reversible drugs which have a number of therapeutic uses, as well as non-competitive, irreversible drugs which are highly toxic. In this practical, the potency of three drugs (neostigmine, tacrine and carbachol) on the activity of brain-solubilized AChE is examined. Whilst neostigmine and tacrine are well-known AChE inhibitors, carbachol is also an inhibitor. Both neostigmine and carbachol contain a carbamyl ester group which tightly binds to the active site of AChE and are actually very slowly hydrolysed. Note that in the experiment described in Section 4.2.3, carbachol was seen to be resistant to the action of both AChE and pseudo- or butrylcholinesterase (pChE). Note that acetylcholine added to the assay decreases the production of thiocholine, although it is not an inhibitor. It does this by substrate completion with acetylthiocholine (ATCh). Thus caution must be taken that a test compound which lowers the rate of thiocholine is a genuine inhibitor by using another AChE assay. The activity of AChE is determined using the continuous visible spectrophotometric method of Ellman *et al.* (1961), in which ATCh is used as a substrate. ATCh is hydrolysed by AChE to thiocholine + acetic acid. Thiocholine formation is detected by reaction with dithiodinitrobenzoate (DTNB) to form 2-nitro,5-thiocholinebenzoate and the yellow product 2-nitro,5-thiobenzoate (TNB) which has an absorption maximum (λ_{max}) at 412 nm. It is the formation of TNB that is actually monitored in this method.

Method Whole rat brain is homogenized (10%) in 30 mM sodium phosphate buffer, pH 7, containing 1% (v/v) Triton X-100 using an Ultra-Turrax homogenizer (setting 6 for 20 s), then centrifuged at 100 000 g for 60 min, and the supernatant stored on ice.

For the *control (uninhibited) assay* pipette into a cuvette:

- 3 mL of 100 mM sodium phosphate buffer pH 8
- 50 μL supernatant into a plastic cuvette
- 100 μL DTNB

Absorbance is monitored by a spectrophotometer connected to a recording device such as PowerLab amplifier and a computer running Chart® software. Alternatively, a kinetic microtiter plate reader can be used. When a stable baseline start is achieved, the reaction is started by adding and mixing 20 µL of 158.5 mM ATCh (final substrate concentration 1 mM). The change in absorbance at 412 nm is monitored for 3 min and recorded and displayed on the monitor by the PowerLab Chart software. This should be repeated three times, since this is the reference for all other tubes, and is needed to calculate % inhibition.

For the *drug-inhibited assay*, make up a series of five logarithmic dilutions (sequential dilutions of 1:10) in distilled water from each of the stock drug solutions (317 µM for neostigmine and tacrine, 317 mM for ACh and CCh) to give a drug concentration range (10^{-5}–10^{-10} M for neostigmine and tacrine, and 10^{-4}–10^{-8} M for ACh and CCh). Do this by adding 100 µL 1 mM stock to 900 µL distilled water, mix and repeat this five times. Be sure to label all tubes with an indelible marker pen. Pipette into a cuvette:

- 2.9 mL of 100 mM sodium phosphate buffer, pH 8
- 50 µL supernatant
- 100 µL DTNB
- 100 µL of each dilution of the inhibitor concentrations

Start the reaction by adding 20 µL ATCh, and mixing the reaction mixture. A linear increase in $E_{412\,nm}$ should be seen, the slope of which is the rate of reaction. If the rate is not linear, ensure that the cuvette has been mixed thoroughly.

Results and Discussion The trace on the computer monitor gives a plot of absorbance at $\lambda = 412$ nm against time (min) for each incubation. The slope (gradient) can be obtained using the Chart® software and this is the enzyme activity with time ($\Delta E/min$). This can be converted to nmol. TNB formed/mg tissue/mL as follows:

$$\text{Activity (nmol TNB/mg tissue/min)}$$
$$= \frac{(\Delta E/min) \times (\text{total. vol. in cuvette}) \times 10^6}{(\varepsilon_{412}\text{TNB}) \times (\text{mg tissue in cuvette})} \quad (8.1)$$

where the molar extinction coefficient ($\varepsilon_{412\,nm}$) of TNB is 1.36×10^4 L/mol/cm. This is the absorbance of a 1 M solution in a 1 cm light path. The total volume in the cuvette is 3.17 mL, and there are $(100 \times 0.05) = 5$ mg tissue in the cuvette. The factor of 10^6 is to convert mL to L.

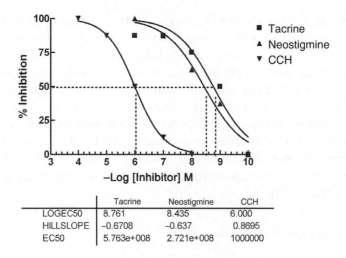

	Tacrine	Neostigmine	CCH
LOGEC50	8.761	8.435	6.000
HILLSLOPE	-0.6708	-0.637	0.8695
EC50	5.763e+008	2.721e+008	1000000

Figure 8.1 Inhibition of AChE by tacrine, neostigmine and CCh. The rates of the enzyme reaction were measured from the slope.

The % inhibition is calculated as:

$$\% \text{ inhibition} = \frac{(\text{control} - \text{inhibited})}{\text{control}} \times 100 \qquad (8.2)$$

Questions

1. Express the drug-inhibited activities as a % of the control and construct semi-logarithmic plots of % control activity against –log [I] (Figure 8.1).
2. Measure the pI_{50} ($-\log IC_{50}$) for each drug tested.
3. Comment on (a) their relative potencies, (b) the mechanism of action of each of the drugs in this assay, (c) possible sources of interference with the method and (d) the suitability of these drugs for particular clinical applications.

8.1 MONOAMINE OXIDASE INHIBITORS

MAO catalyses the deamination of a wide range of substrates including the neurotransmitters noradrenaline, 5-HT and dopamine. This mitochondrial enzyme has a wide tissue distribution, and occurs as two isoenzymes, MAO_A and MAO_B, that are different gene products. MAO_A has a high affinity for the neurotransmitters noradrenaline and 5-HT, and is inhibited by clorgyline. MAO_B is less specific and oxidizes dietary organic amines such as phenylethylamine and benzylamine. It is inhibited by deprenyl. Both enzymes deaminate tyramine and dopamine, and are inhibited by tranylcypromine. MAO can be assayed by several methods including oxygen uptake and radioisotopic measurements. In this practical, a fluorimetric method is used. The deamination of the substrate, kynuramine to the cyclic product 4-hydroxyquinoline (4-OHq), which fluoresces in alkaline solutions, is monitored (Morinan and Garratt, 1985).

Aim This experiment is carried out in two parts:

1. To demonstrate the sub-cellular location of MAO. The relative specific activity (RSA) of MAO in sub-cellular fractions is compared with that of marker enzymes. It is demonstrated that MAO activity is highest in the mitochondrial fraction.
2. To demonstrate the specificity of MAO inhibitors for MAO isoenzymes. Three irreversible inhibitors (tranylcypromine, clorgyline and deprenyl (selegeline)), for MAO_A and MAO_B are tested in an enzyme preparation of whole rat brain.

8.4.1 Sub-cellular Distribution of MAO Activity

The aim of this experiment is to demonstrate that MAO (EC 1.4.3.4) is located in the mitochondrial fraction of a homogenate fractionated by differential and density gradient centrifugal fractionation. A 10% homogenate of rat brain in 0.32 M sucrose/1 mM EDTA, pH 7.2, is prepared using an Ultra-Turrax homogenizer (setting 6 for 20 s). This is centrifuged at 1000 g for 10 min at 4°C to give the nuclear pellet (P1). The supernatant (S1) is carefully removed and centrifuged at 10 000 g for 20 min to give the mitochondrial pellet (P2) and the supernatant (S2). Finally, the microsomal pellet (P3) and the soluble cytoplasmic components (S3) were prepared by centrifugation of S2 at 100 000 g for 60 min. P1 and P2, each resuspended in 15 mL of homogenizing medium together with P3 in 5 mL homogenizing medium and S3, are stored on ice until use in the MAO assays.

MAO Assay of Sub-cellular Fractions

1. Place 16 numbered 1.5 mL Eppendorff microcentrifuge tubes in a rack and pipette 870 μL of 10 mM potassium phosphate buffer, pH 7.2 into each of them. Dispense the following:
2. To tubes 1–4 add 100 μL P1
3. To tubes 5–8 add 100 μL P2
4. To tubes 9–12, add 100 μL P3
5. To tubes 13–16, 100 μL S3.
6. Place all the tubes in a rack and warm them to 37°C for 5 min. Then add 30 μL 3 mM kynuramine substrate (final concentration (92 μM)) to each tube.
7. After 15 min incubation, stop the reaction by adding 30 μL of 0.4 M perchloric acid. Cap and mix the tubes and then centrifuge them at 13 000 rpm. for 1 min in a benchtop microcentrifuge to sediment precipitated protein.
8. Carefully transfer 1 mL of the supernatant from each tube to plastic LP4 tubes and add 2 mL of 1 M NaOH. Mix and read the fluorescence of the product 4-hydroxyquinoline (4-OHq) at $\lambda_{ex} = 305$ nm and $\lambda_{em} = 380$ nm.

4-hydroxyquinoline Standard (calibration) Curve

To calculate the amount of product formed by MAO in the sample tubes, a standard curve is prepared.

1. Dilute the stock 1 mM solution of 4-OHq by 100-fold (in a 25 mL plastic universal white capped tube) by adding 0.1 mL 1 mM 4-OHq to 10 mL 10 mM phosphate buffer. Cap and mix.
2. Pipette reagents (in triplicate) into Eppendorff microfuge tubes: as shown in Table 8.1.

Table 8.1 Preparation of standard concentrations of 4-OHq for the MAO assay.

Tube label	S0	S1	S2	S3	S4	S5	S6	S7
10 μM 4-OHq (μL)	0	50	100	200	400	600	800	1000
10 mM buffer (μL)	1000	950	900	800	600	400	200	0
nmol 4-OHq	0	0.5	1	2	4	6	8	10

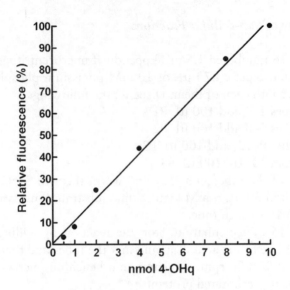

Figure 8.2   Standard curve for 4-hydroxyquinoline

3. Add 300 µL 0.4 M perchloric acid to each tube. Cap and mix.
4. Transfer 1 mL to test tubes containing 2 mL 1 M NaOH. Mix and read the fluorescence at the same wavelengths as used for the sample tubes.
5. Draw a calibration curve of nmol 4-OHq/tube against fluorescence (Figure 8.2).

Protein Assay

This is an adaption of the Lowry method which was first described by Markwell *et al.* (1978). Prepare a stock solution containing 100 µL/mL of bovine serum albumin, and prepare standard tubes containing 0, 10, 20, 30, 40, 50, 60, 70, 80, 90 and 100 µg in 1 mL of distilled water. Make up the following dilutions of the sub-cellular fractions in distilled water: P1 and P2: 1/100 and 1/200; P3 and S3: 1/20 and 1/50. Add and mix 3 mL of Markwell C reagent (1% Reagent B in Reagent A) to duplicate 1 mL aliquots of each of these eight dilutions (a total of 16 tubes) and the standards. After 10 min at room temperature, add with immediate mixing 0.3 mL Reagent D (50% Folin-Ciocalteau Phenol reagent in distilled water). Leave the solutions at room temperature for a further 45 min before reading the absorbance at $\lambda = 660$ nm.

Table 8.2 RSA data for enzymes and nucleic acid markers (Marchbanks, 1975). Fraction P1 is the nuclear fraction, P2, the mitochondrial fraction, P3, the microsomal fraction and S3 is the cytoplasmic fraction. The abbreviations used to denote enzymes are SDH, succinate dehydrogenase, NCcR, NADPH cytochrome c reductase (NADPH cytochrome P450 reductase) and LDH, lactate dehydrogenase.

	% Total marker					RSA			
	Protein	DNA	SDH	NCcR	LDH	DNA	SDH	NCcR	LDH
P1	16	77	8	10	11	**4.8**	0.5	0.6	0.7
P2	41	15	84	21	20	0.4	**2.0**	0.5	0.5
P3	16	5	6	47	8	0.3	0.4	**2.9**	0.5
S3	26	3	2	22	61	0.1	0.1	0.8	**2.3**

Source: Marchbanks, R.M. (1975) Cell-free preparations and subcellular particles from neural tissues, in *Practical Neurochemistry* (ed. H. McIllwain), pp. 208–242. Edinburgh: Churchill-Livingstone.

Calculations

Calculate the RSA of MAO in each sub-cellular fraction. A RSA value greater than 1 indicates an association of the marker with a particular fraction (see Table 8.2) and that the fraction is relatively pure from other organelles. The RSA is defined as:

$$RSA = \frac{\% \text{ total MAO in the fraction (nmol 4-OHq/mL)}}{\% \text{ total protein in the fraction (mg/mL)}}$$

8.4.2 Specificity of MAO Inhibitors for Isoenzymes

For this experiment, a crude homogenate of rat brain can be used. A 10% homogenate of rat brain in 10 mM phosphate buffer, pH 7.2, is prepared using an Ultra-Turrax homogenizer (setting 6 for 20 s). This is then centrifuged at 1000 g for 10 min to remove cell debris and nuclei, and the supernatant carefully removed and stored on ice until required. Acceptable results can be obtained if this homogenate stored at −80°C for several months.

Make seven 10-fold dilutions of the 1 mM stock concentration of each of the inhibitors. Label these as T4 (1 mM stock *undiluted* tranyl-cypromine, final incubation concentration 10^{-4} M), and T5–T11 for subsequent 10-fold dilutions (representing final incubation concentrations of 10^{-5}–10^{-11} M) of tranylcypromine. Follow a similar procedure for clorgyline (C4–C11) and deprenyl (D4–D11). Note that a standard curve for the product of the reaction, 4-OHq, must also be prepared.

This is done as detailed for the cell fractionation experiment (above). This should be done at this stage to prevent any delay at the end of the experiment. Standards should be read with the experimental samples.

1. Place 24 Eppendorff microcentrifuge (1.5 mL) tubes in a rack and label them as follows: three tubes for zero controls (Z1, Z2 and Z3) containing no inhibitor, and one each T4–T11 (for tranyl-cypromine dilutions), C4–C11 (for clorgyline dilutions) and D4–D11 (for deprenyl dilutions).
2. Add 100 μL of homogenate to each tube, followed by 870 μL of 10 mM phosphate buffer to the control tubes and 770 μL buffer to the tubes labelled T, C and D. Add 100 μL of each of dilutions of the inhibitors to the appropriate tubes.
3. Pre-heat for 5 min at 37°C.
4. Start reaction by adding 30 μL of 3.07 mM kynuramine (final concentration 92 μM). Incubate for 15 min at 37°C. Be accurate with the timing.
5. Stop the reaction by adding 300 μL of 0.4 M perchloric acid. Cap and mix and centrifuge at 13 000 rpm in a benchtop microcentrifuge for 1 min to remove precipitated protein.
6. Transfer 1 mL of the supernatant to 4 mL plastic test tubes containing 2 mL of 1 M NaOH. Cap and mix on a vortex mixer.
7. Read the fluorescence of the product, 4-OHq, at $\lambda_{ex} = 315$ nm, $\lambda_{em} = 380$ nm.

Questions

1. Convert the fluorescence readings obtained for each of the sample tubes to nmol 4-OHq formed per 15 min.
2. Calculate the % inhibition obtained in each of the tubes containing inhibitors compared with the control tubes (0% inhibition).
3. Plot –log [MAO inhibitor] M against inhibition of MAO activity (%). See Figure 8.3.
4. From these curves, estimate the IC_{50} (M) values for each of the inhibitors. Some of these inhibitors should give double sigmoidal plots due to their different affinities for MAO_A and MAO_B. Tabulate each of these IC_{50} values.
5. Comment on the specificity of each of the inhibitors, and when appropriate calculate their selectivity for each of the isoenzymes, $MAOI_A$ and MAO_B. On the basis of these results and information from the literature, suggest possible therapeutic applications of these inhibitors and comment on foreseeable adverse reactions.

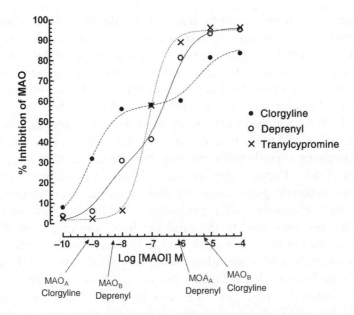

Figure 8.3 Inhibition of MAO by clorgyline, deprenyl and tranylcypromine. According to Morinan and Garratt (1985), clorgyline and deprenyl give biphasic curves from which IC_{50} values for both MAO_A and MAO_B can be obtained. These values show the relative specificity of the inhibitors for the two isoenzymes. (Morinan, A. and Garratt, H.M. (1985) An improved fluorometric assay for brain monoamine oxidase. *J. Pharm. Methods.* 13: 213–223.)

The $IC_{50}(M)$ values reported by Morinan and Garratt (1985) were MAO_A, 3.6×10^{-9} for clorgyline and 3.2×10^{-6} for deprenyl, and for MAO_B, 7.9×10^{-6} for clorgyline and 2×10^{-8} for deprenyl.

8.5 THROMBIN INHIBITORS

Thrombin (factor II) has a central role in controlling haemostasis. Among its many actions, it acts as a procoagulant and anticoagulant factor, as well as fibrinolytic and antifibrinolytic (Siller-Matula *et al.*, 2011). It is serine protease that selectively cleaves the Arg–Gly bonds of fibrinogen to form fibrin, as well as promoting platelet shape change and aggregation. A common method used to assay thrombin activity is by monitoring cleavage of specific chromogenic p nitroanilide sub strates such as phe-pipecolyl-arg-p-nitroanilide (S-2238), benzoyl-Phe-Val-Arg-p-nitroanilide (S-2160) and tosyl-gly-arg-4-nitroanilide (Chromozym TH, Roche Applied Science), all of which result in the liberation of yellow 4-nitroaniline, the appearance of which can be monitored at 405 nm.

There is a major research initiative to develop direct thrombin inhibitors as novel anticoagulant treatments to prevent thrombosis (Tapparelli *et al.*, 1993a; Coppens *et al.*, 2012). Direct thrombin inhibitors have major advantages over other anticoagulants, such as heparins and warfarin. Direct inhibitors bind to thrombin directly and inhibit interaction of thrombin with its substrates. They have three major advantages. They act independently of antithrombin III so they inhibit both free thrombin and fibrin-bound thrombin. They do not bind to plasma proteins giving a predictable anticoagulant effect and they have fewer unwanted side effects. There are three classes of thrombin inhibitors, naturally occurring peptides exemplified by hirudin from leeches, synthetic peptide arginals and peptidomimetics. Hirudins and analogues (hirulogs) not only bind to the catalytic site of thrombin, but also to an apolar and anion binding site and have exceptionally high affinity for thrombin. Synthetic peptidomimetics only bind to the catalytic site. Hirudin and the analogue bivalirudin, as well as some peptidomimetics, are licensed drugs. The latter can be used to prevent thrombosis after hip and knee replacements. During the development of direct thrombin inhibitors, several peptidomimetics proved to have toxic side effects making them unsuitable for clinical applications, but now they are used for laboratory applications. An early class of direct thrombin peptidomimetic inhibitors was the chloromethylketones, of which D-Phe-Pro-Arg-chloromethylketone (PPACK) is an example. PPACK is a high affinity, irreversible thrombin inhibitor that alkylates the amino acid His57 in the active site of the enzyme. Due to the complexities of thrombin activities *in vivo*, it appears to havedifferent affinities in three different assays, the chromogenic assay, platelet aggregation and coagulation (Tapparelli *et al.*, 1993b). The following assay to estimate the apparent affinity of PPACK for thrombin is relatively simple since it is a plasma-free assay which avoids consideration of the effects of other clotting factors on thrombin activity.

Protocol

The assay is carried out using a kinetic plate reader, such as the Thermolabs Labsystem iEMS reader/dispenser MF or the Multiskan (from Thermoscientific), interfaced with a computer running Genesis or SkanIt software. The computer should be programmed such that it can take readings of absorption at 405 nm every 15 min, with 5 s shaking between readings. The assay can be performed manually using a spectrophotometer, but this is more time consuming and some means of recording the time courses of the increase in optical density should be available. An assay buffer is prepared containing 50 mM Tris (pH 8.3), 0.1 M NaCl and 0.1% bovine serum albumin. A 2 mM solution of the substrate (S2160, S2238 or Chromozym TH) is prepared and dispensed into aliquots of 0.2 mL each which should be stored frozen ($-20°C$). A substrate–velocity curve is first constructed to determine the K_m. Serial of dilutions of 1:1 are made in distilled water. Dilute the substrate in a microtiter plate (for later dispensing into a reaction plate) as follows:

Well	Additions	Stock concentration (μM)	Concentration in assay (μM)
A	150 μL 2 mM substrate + 150 μL H$_2$O	2000	200
B	150 μL from A + 150 μL H$_2$O	1000	100
C	150 μL from B + 150 μL H$_2$O	500	50
D	150 μL from C + 150 μL H$_2$O	250	25
E	150 μL from D + 150 μL H$_2$O	125	12.5
F	150 μL from E + 150 μL H$_2$O	62.5	6.25
G	150 μL from F + 150 μL H$_2$O	31.25	3.13
H	150 μL H$_2$O	0.0	–

Bovine thrombin for topical use (Thrombinar) supplied in ampoules containing 1000 IU is a convenient source of standardized thrombin. Before use, add 1 mL distilled water to make a solution of 1000 IU/mL. 50 μL aliquots can be stored in plastic microfuge tubes at –20°C or less. Dilutions are made in the assay buffer. Take 40 μL to a microfuge tube, and add 1 mL of assay buffer to give a solution of 40 IU thrombin/mL. This must be stored on ice.

Estimation of the K_m of the substrate

1. Pipette the following into the wells of a microtiter plate: 160 μL buffer. Add 20 μL of each substrate dilution using a multi channel pipette.
2. The assay is started by adding 20 μL thrombin (40 IU/mL stock) to give a well concentration of 4 IU/mL. An automatic multi-channel pipette should be used to ensure that the reactions start in all the

wells at very nearly the same time. Insert the microtiter plate carrier into the reader and start the reading programme.

3. The time courses should be linear, and most software has some means of calculating the slopes of the enzyme reaction. Plot substrate against velocity (slope) and a classic hyperbolic curve should be produced. An estimate of the K_m can be made more accurately from a secondary plot such as a Lineweaver–Burk, Eadie–Hofstee or Hanes–Woolf plot as described in Section 9.4.3. A value close to 1 μM has been found for all three suggested substrates.

Estimation of the IC_{50} of PPACK

5 mg (9.5 μmol) of PPACK (H-D-Phe-Pro-Arg-chloromethylketone, 2HCl, MW = 523.88) is dissolved in 0.95 mL of distilled water to give a 10 mM solution. 10 μL aliquots can be stored at −20°C. Dilutions of PPACK are made in a 96-well microtiter plate. 150 μL of 10^{-6} M PPACK is placed in a well of row A of the microtiter dish, and 50 μL in row B. To row B is added 100 μL of distilled water (a dilution factor of 1:3) to give a concentration of 3×10^{-7} M. Similar 1:3 dilutions are made down all the rows from Row C to Row G. Row G contains a concentration of 1.3×10^{-9} M. When 20 μL of each of these dilutions of PPACK are added to the reaction dish which contains a total volume of 200 μL, there will be a 1:10 dilution ratio, so final concentrations will range from 1×10^{-7} to 1.3×10^{-10} M in Row A to Row G.

The determination of the K_i of an enzyme inhibitor involves a long experiment because a series of substrate–velocity curves must be constructed in the presence of different concentrations of the inhibitor. For competitive inhibitors, it is possible to obtain an estimate from just one plot in which the IC_{50} is estimated, and if the K_m of the substrate is known then the Cheng–Prusoff equation can be applied. This is shown in Section 8.6 for ouabain, which is an inhibitor of the Na^+,K^+-ATPase. Unfortunately this is not applicable for an irreversible inhibitor, such as PPACK. However, the determination of IC_{50} values are sufficient to give an estimate of the affinity of the inhibitor for an enzyme, and can be compared with IC_{50} values for similar enzymes. In the case of PPACK, it is of great interest to know the specificity for thrombin compared with other serine proteases, such as trypsin.

1. Pipette the following into the wells of a microtiter plate: 140 μL buffer. Add 20 μL substrate at a concentration which produces about 70% of the maximum velocity in the absence of inhibitor (from the Michaelis–Menten curve for the experiment to find the K_m of the substrate). Add 20 μL of each of the PPACK dilutions.

2. The assay is started by adding 20 µL thrombin (40 IU/mL stock) to give a well concentration of 4 IU/mL. An automatic multi-channel pipette should be used to ensure that the reactions start in all the wells at very nearly the same time. Insert the microtiter plate carrier into the reader and start the reading programme.
3. The time courses should be linear, and most software has some means of calculating the slopes of the enzyme reaction. Plot substrate against velocity (slope) and a classic hyperbolic curve should be produced. The negative logarithm of the inhibitor concentration (M) is plotted against the slope (velocity), and the concentration that reduces the velocity by 50% is found. This is the IC_{50}.

This value can be compared by running a similar experiment, but by replacing thrombin with trypsin. A value for the IC_{50} for trypsin should be about 50 times greater than that for thrombin. In addition, an experiment to find the IC_{50} for the inhibition of thrombin-stimulated platelet aggregation can be done by using the microtiter plate-based technique for platelet aggregation described in Section 7.2.

8.6 ATPase INHIBITORS

The Na^+/K^+-activated (Mg^{2+}-dependent) adenosine triphosphatase (ATPase); ATP phosphohydrolase (EC3.6.1.3) is involved in the active transport of monovalent cations and monoamine neurotransmitters across axonal and synaptic membranes. The hydrolysis of one molecule of ATP to ADP and organic phosphate yields 31 kJ of energy which is used to drive the Na^+/K^+ pump and monoamine transporters. In this practical, ATPase activity is measured by visible wavelength spectrophotometry. The inorganic phosphate (HPO_4^{2+} or P_i) formed by the enzyme reacts with ammonium molybdate to give a green-coloured complex (adapted from McNulty et al., 1978). The inhibition of the Na^+/K^+-dependent ATPase by ouabain is examined. This toxic cardiac glycoside is derived from the African plant Strophanthus gratus, and binds to the α-subunit of the enzyme. In this experiment, the IC_{50} and K_i are estimated using the Chung–Prusoff relationship (Chung and Prusoff, 1973).

Method A 10% (w/v) rat brain homogenate in 0.32 M sucrose, 1 mM EDTA, 0.15% (w/v) sodium deoxycholate is centrifuged at 10 000 g for 20 min at 4°C, and the supernatant removed and stored on ice for use in the ATPase assay. The assay is carried out using Tris buffers, with or without the addition of Na^+ and K^+. The Na^+/K^+ buffer contains

50 mM Tris, 10 mM KCl, 5 mM $MgCl_2$ and 150 mM NaCl, pH 7.4 (TKMN buffer), whilst that containing only 50 mM Tris and 5 mM $MgCl_2$, pH 7.4, is abbreviated TM buffer. Phosphate is measured using a molybdate reagent consisting of 1% (w/v) ammonium molybdate + 1% (v/v) Triton X-100 in 1 mM H_2SO_4.

The enzyme assay is carried out in either a conventional spectrophotometer or a microtiter plate reader. The spectrophotometer is coupled to a recording device, such as PowerLab equipment (ADInstruments) and is the same as used for the *in vitro* organ bath experiments. In this case, no preamplifier is used and the voltage output from the spectrophotometer is plugged directly into the analogue input of the PowerLab (AD) amplifier, which in turn is connected to a computer running Chart software and linked to monitor. The changes in voltage are calibrated so that the monitor displays changes in optical density with time. If a microtiter plate reader is used it must be able to record, and preferably display, kinetic readings.

Standard Curve   An inorganic phosphate standard curve is prepared using a series of dilutions of 2 mM Na_2HPO_4 in distilled H_2O ranging from 0 to 1.0 μmol/tube. Each standard is prepared in duplicate in plastic LP4 tubes.

Tube number	2 mM Na_2HPO_4 (mL)	d.H_2O (mL)	μmol Na_2HPO_4
–	0.0	0.5	0.0
S1	0.1	0.4	0.2
S2	0.2	0.3	0.4
S3	0.3	0.2	0.6
S4	0.4	0.1	0.8
S5	0.5	0.0	1.0

Add 2.5 mL ammonium molybdate reagent to each tube, mix and read the absorbance at 405 nm. Plot the phosphate standard curve (μmol phosphate vs. $E_{405 nm}$).

Ouabain Dilutions   Make five 1:10 dilutions of the 10 mM solution of ouabain provided (B1) using microfuge tubes. Place 0.1 mL of B1 into an LP4 tube and add 0.9 mL of distilled water (label this B2). Repeat this procedure to obtain B3–B6.

Incubations   Now triplicate incubations are carried out at 37°C in plastic test tubes in a final volume of 0.5 mL. Controlled (uninhibited) Na^+/K^+ ATPase activity, non-Na^+/K^+-dependent ATPase activity, and ouabain inhibited Na^+/K^+ ATPase activity are determined by pipetting the following volumes of solutions into LP4 plastic test tubes.

Tube number	TKMN buffer (μL)	TM buffer (μL)	Brain supernatant (μL)	Ouabain dilution (μL)
1, 2, 3	400	–	50	–
4, 5, 6	–	400	50	–
7, 8, 9	350	–	50	50 B1
10, 11, 12	350	–	50	50 B2
13, 14, 15	350	–	50	50 B3
16, 17, 18	350	–	50	50 B4
19, 20, 21	350	–	50	50 B5
22, 23, 24	350	–	50	50 B6

Place all the tubes in a 37°C water bath. Start the reaction by adding 50 μL 10 mM ATP. After 15 min, stop the reaction by adding 2.5 mL molybdate reagent (gloves should be worn since this reagent contains H_2SO_4), mix and read the absorbance at 405 nm.

Calculations

1. Calculate the Na^+/K^+-dependent ATPase activity (nmol HPO_4^- produced/mg brain tissue/min) by first subtracting the non-Na^+/K^+-dependent absorbance (tubes 7–9) from that for all other tubes.
2. Calculate the % inhibition caused by ouabain. Display this data on appropriate graphs. See Figure 8.1.
3. Calculate the IC_{50} for the inhibition of Na^+/K^+-ATPase activity by ouabain.
4. Assuming that the K_m of ATP for Na^+/K^+-ATPase is 0.5 mM, calculate the affinity constant (K_i) of ouabain for Na^+/K^+-ATPase

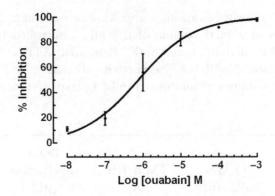

Figure 8.4 Typical results obtained for inhibition of Na^+,K^+-ATPase by ouabain. Each point indicates the mean, and the error bars show the SD, $n = 3$. The curve was fitted to the equation for a sigmoid curve, variable slope using GraphPad Prism. $EC_{50} = 7.48 \times 10^{-7}$ M. (GraphPad Software, Inc, San Diego, CA, USA.)

using the Cheng–Prusoff relationship (Cheng and Prusoff, 1973; Craig, 1993):

$$K_i = \frac{IC_{50}}{1 + (K_m/[S])} \tag{8.3}$$

where $IC_{50} = IC_{50}$ for ouabain (M), $S = [ATP]$ in (M), and K_m is the K_m for ATP (M).

Questions

1. What % of total brain ATPase activity is accounted for by Na^+/K^+-ATPase?
2. What values for IC_{50} and K_i of ouabain for the enzyme did you obtain?
3. Discuss the physiological and pharmacological actions of cardiac glycosides. Quote appropriate references to support your answers.

REFERENCES

Cheng, Y. and Prusoff, W.H. (1973) Relationship between the inhibition constant (K_i) and the concentration of the inhibitor which causes 50% inhibition (I50) of an enzymatic reaction. *Biochem. Pharmacol.* 22: 3099–3108.

Coppens, M., Eikelboom, J.W., Gustafsson, D., Witz, J.I. and Hirsh, J. (2012) Translational success stories: development of direct thrombin inhibitors. *Circ. Res.* 111: 920–929.

Craig, D.A. (1993) The Cheng-Prusoff relationship: something lost in translation. *Trends Pharmacol. Sci.* 14: 89–91.

Ellman, G.L., Courtney, K.D., Andres, V. Jr and Feather-Stone, R.M. (1961) A new and rapid colorimetric determination of acetylcholine esterase. *Biochem. Pharmacol.* 7: 88–95.

Marchbanks, R.M. (1975) Cell-free preparations and subcellular particles from neural tissues, in *Practical Neurochemistry* (ed. H. McIllwain), pp. 208–242. Edinburgh: Churchill-Livingstone.

Markwell, M.A.K., Haas, S.M., Bieber, L.L. and Tolbert, N.E. (1978) A modification of the Lowry procedure to simplify protein determination in membrane and lipoprotein samples. *Anal. Biochem.* 87: 206–210.

McNulty, J., O'Donovan, D.J. and Leonard, B.E. (1978) The acute and chronic effects of D-amphetamine, chlorpromazine, amitriptyline and lithium chloride on adenosine 5-triphosphatases in different regions of the rat brain. *Biochem. Pharmacol.* 27: 1049–1053.

Morinan, A. and Garratt, H.M. (1985) An improved fluorometric assay for brain monoamine oxidase. *J. Pharm. Methods.* 13: 213–223.

Siller-Matula, J.M., Schwameis, M., Blann, A., Mannhalter, C. and Jilma, B. (2011) Thrombin as a multi-functional enzyme. Focus on *in vitro and in vivo* effects. *Thromb. Haemostasis.* 106: 1020–1033.

Tapparelli, C., Matternich, R., Ehrhardt, C. and Cook, N.S. (1993a) Synthetic low molecular weight thrombin inhibitors: molecular design and pharmacological profile. *Trends Pharmacol. Sci.* 14: 366–376.

Tapparelli, C., Metternich, R., Ehrhardt, C., Zurini, M., Claeson, G., Scully, M.F., and Stone, S.R. (1993b) In vitro and in vivo characterization of a neutral boron-containing thrombin inhibitor. *J. Biol. Chem.* 268: 4734–4741.

9

Complementary Methods for Teaching Practical Pharmacology

9.1 THE COMPARATIVE MERITS OF AVAILABLE METHODS

Due to the increasing pressures of increasing class sizes, costs of laboratory experiments and ethical and legal issues, there has been an unremit-
ting search for alternative methods for teaching pharmacology. Available alternatives fall into three major categories. These are computer
aided learning, including software simulation of experiments, problem-
based case studies frequently done in groups and lastly, interpretation of
experimental data. All of these are largely complementary formats to the
teaching of laboratory pharmacology. The challenge is to assess what
types of activities and programs are most suitable and productive. Software for teaching practical pharmacology is widely available and pro-
moted, although not necessarily cheap. In the author's experience, the
most successful types of software are those containing simulated experiments. An obvious advantage of these is that there is no limitation on
the type of animal or preparation that can be used. These are obviously
a good introduction to *in vivo* pharmacology experiments. Typically,
after having obtained an understanding of the principles and aims of
the experiment, some preparation has to be done such as diluting solutions and calibrating machines. Solutions are applied to a preparation or
animal and responses noted on charts or displays. Readings have to

Practical Pharmacology for the Pharmaceutical Sciences, First Edition. D. Michael Salmon.
© 2014 John Wiley & Sons, Ltd. Published 2014 by John Wiley & Sons, Ltd.

be made and tabulated. After appropriate calculations have been performed and graphs prepared, conclusions and explanations are drawn. This clearly is a close approximation to real-life experimentation, but it is complementary and not a complete substitute. Problem-based learning is a useful adjunct to a practical pharmacology course. Computer aided learning and problem-based case studies are well supported by the educational section of the British Pharmacological Society web site (www.bps.ac.uk). Although some say that it is not relevant or of interest for many pharmacology students to do 'hands-on' pharmacology, laboratory science is at the heart of the discipline. There are complaints from some that experiments 'rarely work', but this will frequently be true in learning any complex skill. A lesson soon learnt in the pharmacology laboratory is that a 'typical' response is rarely seen. This is largely due to biological variation of the preparation and lack of experience, and less frequently to idiosyncrasies or malfunctions of the apparatus. One important skill is how to deal with situations when things go wrong.

The majority of this chapter is devoted to the essential task of interpretation of experimental data.

9.2   INTERPRETATION OF EXPERIMENTAL DATA

One of the most useful types of exercise for students to appreciate experiments that are unsuitable to be carried out in most teaching institutions is to process and interpret results obtained from actual experiments. Behavioural pharmacology is one such area as this requires large numbers of animals and specialized facilities, and experiments are time consuming. In addition the acquisition of the appropriate licenses is expensive and may not be justifiable. Other types of experiments are those where expensive equipment and reagents are required, but are commonly used in pharmacology research as carried out in research and development, but it is important to have a grasp of the principles and practice involved, because they are central to pharmacology. Two examples of these are ligand binding and analytical pharmacology. Results of the experiments of these three types are given here, and students are required to process and interpret the results.

9.2.1   Behavioural Experiments

Anxiety is expressed in a wide range of clinical conditions that include phobias and panic disorders. It is viewed as a condition where there is an inhibition of normal behaviour. Although it is a subjective phenomenon in humans, there is a need to attempt to develop some animal models in which to screen potential new drugs for anxiolytic activity.

The exercise here involves assessing the anxiolytic properties of two novel benzodiazepines in two models of anxiety: conflict behaviour and neophobia (elevated plus maze test)

Geller–Seifter Conflict Tests

This test involves the use of an operant conditioning chamber or Skinner box which contains a lever that a small rodent can press in order to obtain food. Rats were deprived of food for 24 h and then allowed to acclimatise in the box for 10 min. In the first 5 min period, each lever press resulted in the delivery of one 45 g food pellet, but in the second 5 min period, each lever press resulted in the delivery of food pellet simultaneously with a small electric current passed through small metal rods on the floor of the box. This is termed the conflict-to-obtain-food or avoid an electric shock. Rats accommodated in this way are termed conflict-trained. Ten groups of 10 rats each were treated as shown in Table 9.1. Drugs were dissolved in sesame oil (sub-cutaneous), and 30 min before the trial, control rats (CON) were given sesame oil only (0.1 mL/kg body weight). All experiments were carried out under the Animals (Scientific Procedures) Act, 1986.

Table 9.1 Geller–Seifter Conflict Test Data. Each value represents the median number of lever presses in the 5 min block (reward or conflict) for 10 rats.

Drug	Dose	Reward	Conflict
Control	–	295	30
MBX1502	0.3 mg/kg	285	40
	1.0	300	100*
	3.0	310	190*
	10.0	320	200*
MBX3043	3 µg/kg	300	75*
	10	293	170*
	30	289	210*
	100	55*	10*
Diazepam	2 mg/kg	290	165*

*$P < 0.05$ compared with control.

Table 9.2   Elevated plus maze data. Each value represents the median time (s) spent in the two open arms or median total time of arm entries over 5 min, for $n = 10$ rats.

Drug	Dose	Time in open arms (s)	Total number of arm entries
Control	–	55	17
MBX1502	0.3 mg/kg	61	16
	1.0	95*	22*
	3.0	125*	26*
	10	140*	30*
MBX3042	3 μg/kg	93*	18
	10	125*	17
	30	160*	19
	100	15*	5*
Diazepam	2 mg/kg	148*	19

*$P < 0.05$ compared with control.

Elevated Plus Maze (Neophobia) Test

The test apparatus consists of two open arms (45 cm×15 cm) and two enclosed arms at 180° to each other. The entire maze was made out of 5 mm thick black perspex and raised 50 cm above the ground. Each rat was given a 5 min trial, and the time spent in the open arms and total (open + enclosed) number of arm entries were recorded (Table 9.2). Injections were administered as for the Geller–Seifter conflict test above.

Nociception (Hot Plate) and Locomotor Activity Tests

To check for false positives in the conflict and neophobia tests, the effects of the novel drugs on nociception (using morphine as a positive control) and spontaneous locomotor activity (LMA) (using amphetamine as positive control) were assessed. To test for any anti-nociceptive effects, rats were placed on a metal plate maintained at a temperature of 56°C and the time taken to lift their forepaws off the plate (reaction time) was recorded. To measure LMA, rats were placed in a darkened box (45×30×30 cm) with parallel rows of photocells placed 4 cm above the flaw of the box. Activity counts (photocell beam interruptions) were recorded over a 5 min period (Table 9.3). All drugs were administered 30 min (s.c.) before the experiment, as for the other tests.

Table 9.3 Nociception and LMA test data. Each value represents the median reaction time (s) or median number of activity counts measured over 5 min for $n = 10$.

Drug	Dose	Reaction time	Activity counts
Control	–	3	450
MBX1502	0.3 mg/kg	2	475
	1.0	12*	525*
	3.0	15*	560*
	10	18*	590*
MBX3042	3 µg/kg	3	454
	10	3	462
	30	5	439
	100	4	45*
Morphine	15 mg/kg	21*	n.d.
Amphetamine	2.5 mg/kg	n.d.	606*

*$P < 0.05$ compared with control.
n.d., not determined.

Questions

Anxiolytic Drug Data Interpretation

1. Do the data in Tables 9.1 and 9.2 suggest that MBX1502 and/or MBX3042 have anxiolytic properties?
2. Which of the two drugs has
   (a) the more potent anxiolytic action,
   (b) a sedative effect,
   (c) an anti-nociceptive effect and
   (d) a stimulant effect?
3. Diazepam (Tables 9.1 and 9.2), like amphetamine and morphine (Table 9.3), may be described as a positive control. Briefly explain the importance of including positive controls in these experiments, and the difference between a positive control and the control (CON, drug vehicle or solvent).
4. What types of drugs could give false positives in the conflict and neophobia test? How do the data in Table 9.3 clarify whether the anxiolytic effects of MBX1502 and MBX3042 are real or secondary to some other pharmacological property of these drugs?
5. Another method to measure anxiolysis is the food preference (hypo neophagia) test. Here food-deprived rats are given a 10 min trial during which they have a choice between eating familiar laboratory food pellets and a novel food (e.g. chocolate buttons). What effect would you expect an anxiolyic drug to have in this test compared to a drug that has appetite-supressant properties, but not on anxiety?

9.2.2   Analysis of Metabolites of 5-hydroxytryptamine

The action of drugs interfering with the metabolism of 5-hydroxytryptamine (5-HT) in the brain are of interest because alterations in 5-HT have been implicated in several altered behavioural conditions, such as affective disorders, anxiety disorders and schizophrenia. The precursor of 5-HT is dietary L-tryptophan that is converted to 5-HT by tryptophan hydroxylase and dopa decarboxylase. 5-HT is degraded to 5-hydroxindoleacetic acid (5-HIAA) by the enzymes monoamine oxidase and aldehyde dehydrogenase. These metabolic products can be separated by chromatography. In the following exercise you are required to analyse and interpret the actions of drugs on 5-HT metabolism.

The following data was adapted from Reinhard *et al.* (1980). Three groups of Lister Hooded rats were injected with 75 mg/kg pargyline at 1.5 h, 2.5 mg/kg at 4 h or saline (control) at 1.5 h, before the whole brain was removed and homogenized in 10 volumes of ice-cold 0.2 M $HClO_4$. Following centrifugation, at $10\ 000 \times g$ for 20 min at 4°C, and the supernatant was passed through a 0.45 μm filter to remove particulates before measuring the concentrations of 5-HT and its metabolite 5-HIAA (5-hydroxyindole 3-acetic acid) using HPLC-ECD (high performance liquid chromatography with electrochemical detection), as described below.

Fifty microlitres of aliquots of supernatant (containing 250 μg protein) were injected via a 50×2.1 mm guard column onto a 300×3.9 mm 5 μm C-18 RP (reverse phase) column. The flow rate of the mobile phase (6% v/v methanol in 0.1 M sodium acetate buffer, pH 4.7; filtered and degassed immediately before use) was set to 2.0 mL/min and the potential of the ECD to 0.7 V. Under these conditions, the retention time of 5-HT was 9 min and for 5-HIAA 12 min. A range of standard indoles (0–40 pmol) made up in 0.2 M $HClO_4$ (Table 9.4), was

Table 9.4   ECD responses for 5-HT and 5-HIAA standards.

	ECD response (nA)	
Indole (pmol)	5-HT	5-HIAA
0	0.00	0.00
5	0.55	0.40
10	1.15	0.85
20	2.25	1.65
30	3.40	2.50
40	4.50	3.30

Table 9.5 ECD responses for 5-HT and 5-HIAA in brain samples from three groups of rats. Group A were injected with pargyline (75 mg/kg), group B were injected with reserpine (2.5 mg/kg) and group C were injected with saline only.

Group A		Group B		Group C	
5-HT (nA)	5-HIAA (nA)	5-HT (nA)	5-HIAA (nA)	5-HT (nA)	5-HIAA (nA)
2.95	0.28	1.75	0.92	0.68	1.3
3.20	0.24	1.65	0.88	0.65	1.32
3.10	0.23	1.68	0.87	0.72	0.29
2.90	0.25	1.70	0.91	0.73	1.31

used to quantify 5-HT and 5-HIAA concentrations in the brain samples (Table 9.5).

9.2.3 Radioligand Binding

Techniques for measuring the direct binding of drugs (usually antagonists, more accurately termed 'ligands') to receptors were first introduced in the 1970s. If the simplest model is assumed, then

$$\text{Ligand} + \text{Receptor} \underset{k_{-1}}{\overset{k_{+1}}{\rightleftharpoons}} \text{Ligand–receptor complex} \tag{9.1}$$
$$\text{L} \quad + \quad \text{R} \qquad \qquad \qquad \text{LR}$$

According to the Law of mass action,

$$K_d = \frac{k_{-1}}{k_{+1}} = \frac{[L][R]}{[LR]} \tag{9.2}$$

where K_d = equilibrium dissociation constant, and k_{+1} and k_{-1} are the association and dissociation and reverse rate constants.

The aim is to determine the affinity of the ligand for the receptor, which is defined as K_d. It is necessary to determine the amount of ligand in the free form [L], and the amount in the bound form [LR]. The earliest techniques involved using a ligand labelled with a radioactive isotope, usually ^{3}H, ^{14}C, ^{125}I, ^{32}P or ^{35}S. The radioactively labelled ligand bound to the receptor to form a ligand–receptor complex. When equilibrium had been reached, the bound ligand was separated from the free ligand, frequently, by filtration through glass-fibre filters or centrifugation.

More recently, other techniques to measure receptor binding have been introduced. Scintillation proximity assays using radioactively labelled ligands can detect bound ligand without separation from the free ligand. Currently, non-radioactive assays based on the detection of fluorescence or chemoluminescence are most commonly used.

Whichever method is used, some measure of free and bound ligand must be obtained. If the amount of bound ligand found on the filter, [LR], is plotted against the total concentration of ligand pipette into incubation (i.e. [L] + [RL]), a hyperbolic curve that reaches a maximum or plateau, should theoretically be found. In fact, a curve that shows no sign of reaching a plateau results. This is due to a complication of radioactive ligand-binding experiments in that there is a portion of the labelled ligand that binds to low affinity, high-capacity sites other than the receptor, representing binding to the containers, which is termed non-specific binding. This can be measured by carrying out the binding experiment in the presence of saturating concentration of non-radioactive ligand, which displaces radioactive ligand from the receptor sites. The amount of 'true' specifically bound ligand is found by first subtracting the non-specific binding from total binding as depicted graphically in Figure 9.1.

The specifically bound curve now reaches a plateau, which is the maximum amount of ligand that can bind to the receptors, meaning that the receptor bonding sites are saturated and there is 100% occupancy of

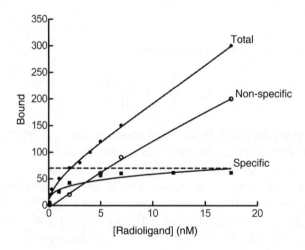

Figure 9.1 Total binding, non-specific and specific binding plotted against concentration. Specific binding is the total minus the non-specific binding.

receptor sites. This is B_{max}. The K_d is theoretically equal to the concentration of ligand that gives 50% receptor occupancy ($B_{max}/2$). However, it is more accurate to calculate K_d from a graph of a linear transformation of the equation. Such graphs can be derived from several types of algebraic rearrangements. The most popular transformation is the Scatchard plot, where according to the equation

$$\frac{B}{F} = -\frac{B}{K_d} + \frac{B_{max}}{K_d} \qquad (9.3)$$

where B = bound, F = free, K_d = dissociation constant and B_{max} = total binding sites, if B (x-axis) is plotted against B/F (y-axis), a straight line should be obtained, where the slope then equals $-1/K_d$, the intercept on the y-axis is B_{max}/K_d, and the intercept with the x-axis = B_{max}. As when analysing enzyme kinetics, there is frequently a deviation from linearity that will indicate some complicating factor in the analysis. Scatchard analysis is very sensitive to such deviations from the simple one-site binding model. For example, a concave curve can indicate negative co-operativity in the interaction between the ligand and the binding site, or it may indicate multiple binding sites on the receptor. A biphasic curve, first rising and then falling, can indicate positive co-operativity, or that the ligand is breaking down during the assay. Several other transformations, familiar to students of enzyme kinetics, such as the Eadie–Hofstee and Hanes–Woolf plots, have also been used. A Hill plot can be used to test whether the interaction between the receptor and the ligand is competitive. This involves plotting the logarithm (bound/B_{max} − bound) against the log[ligand]. More recently, it has become more clear that such plots are subject to some statistical flaws, and it is better to perform computer-fitting of binding data to an equation.

Ligand-binding Exercise

Rat hippocampal synaptic plasma membranes (equivalent to 20 mg wet weight) isolated by differential centrifugation, were incubated at 37°C with [G-^{3}H]-5-hydroxytryptamine (5-HT: 1.2–36.0 nmol/L, *specific radioactivity*: 18.0 Ci/mmol) in a final volume of 2 mL PIA buffer pH 7.6 (50 mmol/L sodium phosphate, 1 μmol/L iproniazid, 10 mmol/L ascorbate). A parallel set of incubations containing in addition, 10 μmol/L 8-hydroxy-2-(di-n-propylamino)tetralin (8-OH-DPAT) a 5-HT$_{1A}$ receptor agonist, was used to correct for non-specific binding. After a 10 min incubation time, the reactions were terminated by rapid filtration under

Table 9.6   Binding of [G-^{3}H]-5-HT to 5-HT$_{1A}$ receptors in the hippocampus. Values for total binding and non-specific binding are the means of $n = 4$. 1 Curie (Ci) $= 2.22 \times 10^{12}$ dpm (disintegrations per minute).

[G-^{3}H]-5-HT (nmol/L)	Total binding (dpm)	Non-specific binding (dpm)
1.2	2264	535
2.4	3680	902
3.6	5162	1200
4.8	6517	1735
7.2	8824	2540
18.0	16 610	7264
36.0	26 405	15 245

vacuum through 2.4 cm GF/B filters. The filters were then washed twice with 5 mL ice-cold PIA buffer and transferred to plastic vials for liquid scintillation counting. The results for both sets of incubations are presented in Table 9.6.

Questions

Calculate the affinity (K$_d$, nmol/L) of [G-^{3}H]-5-HT for the 5-HT$_{1A}$ receptor and the total number (B$_{max}$ pmol [G-^{3}H]-5-HT bound/g tissue) of 5-HT$_{1A}$ receptors in the hippocampus, from the following linear plots:

1. Scatchard (B/F v B); m $= -1/$K$_D$, c $=$ B$_{max}$/K$_d$ *when* y $= 0$, x $=$ B$_{max}$ as shown in Figure 9.3.
2. Eadie–Hofstee (B v B/F); m $= -$K$_d$ c $=$ B$_{max}$ as   shown   in Figure 9.4.
3. Hanes–Woolf (F/B v F); m $= 1/$B$_{max}$ c $=$ K$_d$/B$_{max}$ *when* y $= 0$, x $= -$K$_d$ as shown in Figure 9.5.
   Fit a line by linear regression. Calculate K$_d$ and B$_{max}$ from the equation for the line.
4. Calculate the mean (μ) $\pm$ SD for K$_d$ and B$_{max}$ from the three plots.
5. To what type of effector is the 5-HT$_{1A}$ receptor linked? Give one other example of this class of receptor and another 5-HT receptor sub-type coupled to a different type of effector.
6. What is the clinical indication for the 5-HT$_{1A}$ receptor ligand, buspirone? Give one other example of a drug that is used for this disorder.
7. Why were iproniazid and ascorbate included in the incubation buffer?

ANSWERS TO QUESTIONS

Anxiolytic Drug Data Interpretation

1. Do the data in Tables 9.1 and 9.2 suggest that MBX1502 and/or MBX3042 have anxiolytic properties?

   Answer: From Table 9.1, both drugs appear to have anxiolytic properties as they both increased the scores in the conflict component of the test, as does diazepam. Also, in the elevated plus maze test, both drugs suggested anxiolytic activity since the rats spent more time in the open arms. However, MBX3042 resulted in an increased total number of entries, which is characteristic of generalized increased activity, and is not seen in response to diazepam.

2. Which of the two drugs has the more potent (a), anxiolytic action, (b) sedative effect, (c), an anti-nociceptive effect and (d) a stimulant effect?

   Answer: (a) The data in Tables 9.1 and 9.2 together suggest that MBX3043 has the greater anxiolytic activity. (b) From the decrease in reaction time in Table 9.3, MBX has the greater sedative effect. (c) MBX1502 has the greater anti-nociceptive effect since, like morphine, there was a significant increase in reaction time in Table 9.3. (d) MBX1502 showed the greater stimulant activity, both in increased activity counts in the LMA test (resembling amphetamine), and the increased number of total arm entries in the elevated plus maze test.

3. Diazepam (Tables 9.1 and 9.2), like amphetamine and morphine (Table 9.3), may be described as a positive control. Briefly explain the importance of including positive controls in these experiments, and the difference between a positive control and the control (CON, drug vehicle or solvent).

   Answer: The positive control is required as evidence that the test is functioning in the expected manner and giving predictable responses. Here diazepam, morphine and amphetamine are used as positive controls. The vehicle only control is included as evidence that the vehicle does not produce responses on its own, and responses are solely due to the drug.

4. What types of drugs could give false positives in the conflict and neophobia test? How do the data in Table 9.3 clarify whether the anxiolytic effects of MBX1502 and MBX3042 are real or secondary to some other pharmacological property of these drugs?

Answer: Stimulant drugs, such as amphetamine could produce false positives in the test for anxiolytic activity. Although MBX1502 resembled diazepam in the responses seen in the conflict and neophobia tests, it is clear from the responses in the LMA test that MBX1502 has stimulant activity.

5. Another method to measure anxiolysis is the food preference (hypo-neophagia) test. Here food-deprived rats are given a 10 min trial during which they have a choice between eating familiar laboratory food pellets and a novel food (e.g. chocolate buttons). What effect would you expect an anxiolytic drug to have in this test compared to a drug that has appetite-suppressant properties, but not on anxiety?

Answer: Anxiolytic drugs such as diazepam usually cause an increase in appetite for unfamiliar food they could easily distinguish from appetite suppressant drugs.

Analysis of Metabolites of 5-hydroxytryptamine

1. Draw fully labelled calibration (standard) curves of ECD response (nA) against amount (pmol) of 5-HT and 5-HIAA.

Answer: The standard curve is constructed by plotting the electro-chemical detector (ECD) response (nA) against the indole concentration (pM) as shown in Figure 9.2.

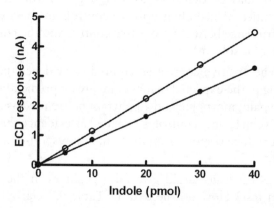

Figure 9.2   Standard curve for the ECD response against indole concentration.

2. Use the equation $y = mx + c$ to calculate the amount of 5-HT and 5-HIAA in the brains of each group of rats.

3. Calculate the average concentration (pmol/mg protein) of 5-HT and 5-HIAA in the brains of each group of rats. See Table 9.7.

Answer:

Table 9.7 Mean concentrations (pmol/mg) ± SD of 5-HT and 5-HIAA in the three groups of treated rats.

	Saline	Pargyline	Reserpine
5-HT	6.15 ± 0.32	26.92 ± 1.22	15.02 ± 0.37
5-HIAA	15.75 ± 0.16	3.00 ± 0.26	10.79 ± 0.29

4. Identify, with reasons which drug treatment each group of rats has received.

Answer: Pargyline is a monoamine oxidase inhibitor that prevents the domination of 5-HT, and 5-HIAA is the final degradation product of 5-HT. Thus pargyline causes a large increase in 5-HT concentration and a five-fold fall in 5-HIAA. Reserpine blocks the vesicular monoamine transporter, VMAT, that transports the monoamine neurotransmitters noradrenaline, dopamine and 5-HT from the cytoplasm of the presynaptic nerve terminal into the storage vesicles. Thus when the transporter is blocked by reserpine, there is a build-up of 5-HT in the cytoplasm, and a moderate decrease in 5-HIAA.

5. What would be the effect on 5-HT concentrations of administering (a) L-tryptophan + carbidopa, and (b) p-chlorophenylalanine?

(a) Carbidopa is an inhibitor of aromatic L-amino acid oxidase which is required for the conversion of L-tryptophan to 5-HT. Tryptophan on its own would increase 5-HT concentrations. If they were administered together there would be little net effect on 5-HT concentrations, but it depends on the relative doses of the two substances. Aromatic L-amino acid oxidase is not a rate-limiting enzyme in the synthesis of 5-HT, so this effect could be overcome by high doses of L-tryptophan.

(b) *Answer*: p-chlorophenylalanine is a specific inhibitor of the rate-limiting enzyme in 5-HT biosynthesis, tryptophan hydroxylase. Consequently, it depletes concentrations of brain 5-HT.

Radioligand Binding

Calculate the affinity (K_d or K_d nmol/L) of $[G-^3H]$-5-HT for the $5-HT_{1A}$ receptor and the total number (B_{max} pmol $[G-^3H]$-5-HT bound/g tissue) of $5-HT_{1A}$ receptors in the hippocampus, from the following linear plots:

1. Scatchard (B/F v B); $m = -1/K_d$, $c = B_{max}/K_d$ *when* $y = 0$, $x = B_{max}$

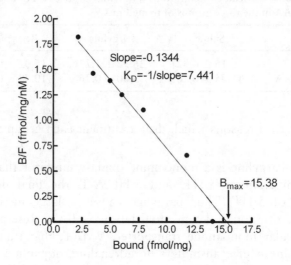

Figure 9.3 Scatchard plot for 5-HT-binding data.

2. Eadie–Hofstee (B v B/F); $m = -K_d$, $c = B_{max}$

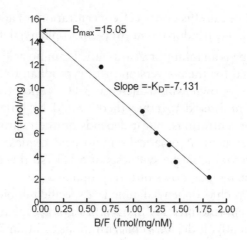

Figure 9.4 Eadie–Hofstee plot for 5-HT-binding data.

3. Hanes–Woolf (F/B v F); m $=$ $1/B_{max}$, c $=$ K_d/B_{max} *when* $y = 0$, $x = -K_d$

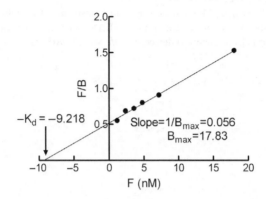

Figure 9.5   Hanes–Woolf plot for 5-HT-binding data.

4. Calculate the mean (μ) $\pm$ SD for K_d and B_{max} from the three plots.

Answer: The mean K_d was 7.93 $\pm$ 1.13 nmol/L and mean B_{max} was 16.44 $\pm$ 1.39 fmol/mg.

5. To what type of effector is the 5-HT$_{1A}$ receptor linked? Give one other example of this class of receptor and another 5-HT receptor sub-type coupled to a different type of effector.

Answer: The 5-HT1A receptor is linked through G_i/G_o to the inhibition of cyclic AMP accumulation and inhibits further release of both 5-HT and noradrenaline.

6. What is the clinical indication for the 5-HT$_{1A}$ receptor ligand, buspirone? Give one other example of a drug that is used for this disorder.

Answer: Buspirone is a newer treatment for generalized anxiety. Benzodiazepines are the older treatment, but the mechanisms of action are completely different.

7. Why were iproniazid and ascorbate included in the incubation buffer?

Answer: Iproniazide is a monoamine oxidase inhibitor. This would prevent degradation of the labelled 5-HT during the binding experiment. Vitamin C (ascorbate) is added to prevent oxidation of 5-HT.

REFERENCES

Animals (Scientific Procedures) Act, 1986 and Amendments (2012) www.legislation.
gov.uk/ukdsi/2012 (accessed on 3 June 2013).

Reinhard, J.F., Moscowitz, M.A., Sved, A.F., and Fernstrom, J.D. (1980) A simple, sensitive and reliable assay for serotonin and 5-HIAA in brain tissue using liquid chromatography with electrochemical detection. *Life Sci.* 27: 905–911.

10

Communicating Results

All the time, effort and expense invested in performing an experiment is lost if the results and conclusions are not communicated to supervisors, peers and the wider scientific community. This is usually done in stages to allow for a series of criticisms and improvements. The first stage is carried out in the laboratory book, and as mentioned in Section 1.2, involves the tabulation of the primary or raw data. In order to assess any relationships between the variables, secondary calculations are made and the results presented in tables or graphs, and statistical tests applied to assess the probability of the variables conforming to these relationships. They are then communicated to peers, such as fellow students or co-workers, for criticism. This is done either orally or in the form of a poster presentation. The next stage is to write a formal report or scientific paper. Because a report is regarded as preliminary there is no excuse to ignore correct scientific units, nomenclature and conventions. This is done so as to avoid ambiguity and to communicate with greater clarity.

10.1 PRELIMINARY REPORTS

Before performing any secondary calculations, often it is convenient to reduce very small or large numbers by converting them to more manageable numbers. This also helps to avoid errors incurred when handling numbers to the power of 10. This can be done by expressing numbers as prefix, for example by expressing 10^{-8}g as 10 ng. However you should be aware that this cannot always be done, for example when drawing a Schild plot to find a pA_2 value, as the pA_2 is defined in terms of molar

Practical Pharmacology for the Pharmaceutical Sciences, First Edition. D. Michael Salmon.
© 2014 John Wiley & Sons, Ltd. Published 2014 by John Wiley & Sons, Ltd.

concentration, although it does not have any units itself. The next stage is to decide on which is the best way to display the data – be it tabular or graphical. Frequently, the type of presentation is obvious, such as constructing a log concentration–response curve. In many cases, the form in which data are presented may not seem so clear, but this is a relatively simple procedure. There are two major forms of data presentation, tables and graphs. Data should always be expressed as means ± an error or (SD or SEM), together with the number of observations or samples. A single observation has little value!

10.1.1   Tables

Tables are only used to report descriptive data. For example, they may be used to show data about the subjects used in a study. Large tables of raw data consisting of individual readings from a machine, such as a spectrophotometer, should not be included in the text. They may be included in an appendix or even only stored in the laboratory notebook. The central point of data presentation is to clearly summarize a large number of individual data in a form that helps the reader understand the data. If it is intended to convey any kind of relationship between the data sets, then a table is not the right format to choose.

10.1.2   Graphs

It is important to select the most appropriate format of graphical format. There are different types of graphs: x–y, bar graphs and histograms and pie charts. An x–y graph is selected if it is desired to demonstrate a relationship between two variables. It is vital to select which variables are the dependent and the independent. Parameters such as time, dose or age are independent variables and are plotted on the x-axis. Responses are dependent variables and are plotted on the y-axis. It is important to pay attention to several details:

- The use of appropriate scales. Should the range of values on the axis begin at zero, or should the range cover only those of the experimental values? If there is a linear axis, the range of values should start at zero (unless there are both positive and negative values, in which case, the range should cover both positive and negative numbers, but the line of the other axis should pass through zero). If the values on the axis are logarithmic numbers, the range of values should only cover the range of the experimental data. Remember that zero on a logarithmic axis gives an original (antilog) value of 1.

- The correct labelling of axes. Each axis should be labelled with a descriptive label together with units. For example, 'enzyme activity (ΔE/min)'. Each axis contains major and minor division marks. Sufficient minor division marks should be included so as to be able to read off values from the line, but not so many as to make the axis very crowded.

- Every graph should be labelled with a figure number, together with a fully descriptive legend, which is generally placed below the graph. If the graph is prepared using a graphics program, the legend should not be written on the graph within this program. The legend should be written within the word-processing program (such as Word). This should contain basic details of the technique used and the variables measured. Identification of symbols used for each line must be included. A brief headline title is not used. The general rule is that a figure should be intelligible without reference to the text.

10.1.3 Bar Graphs

These are used when there is a dependent variable and a discontinuous variable, such as treatments or types. Again, for the independent variable there must be a descriptive label and correct units. Bar graphs and histograms must be labelled with a figure number and a descriptive legend, as for x–y graphs. Rarely, a pie chart is a clear way to describe data. These are best used to show the distribution of a dependent variable in terms of abundance (as a percentage) amongst different groups or treatments.

10.1.4 Preliminary Conclusions

After the data have been plotted, and a relationship or statistically significant differences established, general conclusions can be drawn. This will form the basis of the abstract in a full report of the work. This should take the form of several brief sentences making distinct points, and may be numbered. First, a brief description of the hypothesis and aims should be stated. This is followed by a list of the results, stating important values together with SD or SEM. Finally, state what may be concluded from the study together with any limitations.

10.2 POSTER PRESENTATIONS

Poster presentations are commonly used as the first public presentation of research work. This may be a preliminary assessment of student work or more widely at scientific meetings. Some scientific societies, such as the British Pharmacological Society (BPS) and Physiological Society, actually perform an assessment after the session and a vote is taken as to whether they are acceptable to the society. A central feature of posters is that they are held in a large space with many other presentations, and they are competing for attention. They are rather like an advertisement for the work. A balance has to be struck between making the presentation attractive and easily assimilated, yet including sufficient details to make them scientifically valid. Authors must be in attendance at the poster, prepared to give further explanation and clarification. They are in a position rather like a shop assistant. They must be available for assistance, yet not so intrusive that they prevent a reader from examining the work.As with all presentations, written or oral, it is vital to be direct it to the type of audience. Frequently, readers of a poster will not be familiar with the topic in detail, and come from a wide range of backgrounds. The actual form of the poster must conform to requirements distributed by the organizers, who should specify the size of the display area, and the type of board and other facilities that will be available to authors. Posters have a title area at the top and the space below is divided into a series of sections. The poster may be prepared as one large sheet that can be rolled up and carried in a cardboard tube, or as separate cardboard sections. Whichever way is chosen, they are best prepared using specialized software, such as Microsoft PowerPoint or Publisher. Large sheets can be printed using an A2 or A1 printer.

The title should be in a large font (such as 72 point) across the top, such that it can be read across a room. It should contain authors' names and affiliations. The wording of the title is important as it has to attract people from a distance. It should be short and attractive, rather like a newspaper headline, yet convey sufficient meaning of the contents of the poster.

From the start, a font and style should be chosen so that it may be easily read from a distance of about 1 m. Text should be double spaced. This means that the amount of details that can be included is limited. The arrangement of the individual sections varies, but it must be clear to follow and be easy to understand how the different sections interrelate. The sections resemble a report, but may not necessarily be as formal as a report. For example, the 'Introduction' may be labelled 'Background'. Since readers have little time to assimilate the contents of a poster, they may be drawn to the end section or 'Conclusions' to find if this is of interest to them. For this reason, it can be a good idea to place a section entitled 'Summary' at the beginning underneath the title. The methods section presents a challenge as it must briefly indicate how the experiments or studies were carried out. The 'Results' are important as readers will want to know what the study has contributed to the topic. It is best to limit the amount of text by including two or three carefully selected graphs, diagrams or photographs. Well-presented visual aids are more rapidly assimilated by readers. The poster should finish with a brief conclusion, limited to a 'take away message'. It is useful to cite one or two key references. It is common for an author to have a small A4-sized copy of the poster that can be given to especially interested readers. Supplementary material can also be available to aid with explanations.

Finally, posters are judged not only by the clarity with which information is conveyed by the poster itself, but also by how the author responds to questions and requests for further explanation. It is important that the author responds to the best of his knowledge, but it is crucial that the author does not go beyond this and extrapolate or invent answers. Equally, if few questions are forthcoming, this is not necessarily a sign of lack of interest; it may be merely a reflection of how clearly the poster was presented. When a poster is presented to a professional society, such as the BPS or the Physiological Society, sometimes there is a vote by the members to ascertain whether the poster presentation is of a standard acceptable to the society. If accepted, an abstract of the work will be published in the proceedings of the society.

10.3 ORAL PRESENTATIONS

Presentations in front of an audience naturally are a cause for nervousness. This is only overcome by having a good familiarity with the topic, as well as background material. The first step in planning a talk is to create a plan with the allotted time in mind. A simple structure is important so that a coherent thread is evident. No extraneous or irrelevant material should be included as this only confuses, and can annoy, listeners. It will also keep the speaker on track. The aim is to become so familiar with the material that there is time to interact with the audience, if this is the only time to look around. With the plan in mind, appropriate slides can now be selected, or if necessary specially created for the purpose. It can be confusing and off-putting to include slides or results that were designed for another purpose. It is best not to include too much detail on each slide, and the aim is not to read the entire talk from the slides, as this will not instil an air of confidence either in the speaker or the audience. The text on slides should only be a cue, and require further explanation, so that the audience listens to the speaker, rather than reading slides. It is vital that the font must be large enough to be clearly read from the back of the room. A simple slide layout is most suitable, and background colours add little to enhance a talk, especially if it makes the text less legible. It is not a good idea to include too many gimmicky photographs and animation, and the design of slides should be kept simple. It has been many years now since the introduction of PowerPoint, and people are becoming tired of being sidetracked by all of its curious features. Remember that it is an aid to explain the study and not an opportunity to advertize the software. When presenting data, avoid complex slides containing too many numbers, lines on graphs or bar graphs. It is best to only present summary statistics, and modify graphs to reduce the amount of data. More detailed slides can be available that could be used to illustrate possible questions at the end of the talk. The aim is to keep control of the talk so that it can be delivered in a paced manner, and not a rushed reading of an excessive number of slides.

Having arrived at a plan of the talk and selected the appropriate slides, it is vital to have complete familiarity with the structure and length of the presentation. This can only really be done by delivering it orally. What appeared a good idea written down can become cumbersome when it is spoken. The order of presentation of slides can be changed, or any vital missing pieces included. The talk then should be practised repeatedly until this can be done smoothly without reference to notes.

It is almost obligatory to answer questions at the end of the talk. As was mentioned in connection with poster presentations, answers should be confined to knowledge that is familiar to the speaker, no more. If the answer to a question is not known, this is no cause for shame and should be answered as such. Some questions are predictable and can be planned for. For example, it is common to ask the speaker what future studies should be done, or how could the data be improved.

10.4 PROJECT REPORTS

Scientific writing can present quite a challenge to many students. It is an exercise in clear planning and coherent expression of the material. It is a well-known weakness, which is not unique to students, to plagiarize material from the internet. With the introduction of software to detect plagiarism, it is not uncommon for students to spend hours rephrasing paragraphs in a vain attempt to avoid detection by such tools. With some thought, it is possible to write original text in less time. The secret to report writing is planning. This is done in a variety of ways that suits different people. A simple approach to planning a project report is to write the contents page first. The basic structure of a scientific report has a conventional form – introduction, methods, results, discussion. If required, an appendix can be added. Sometimes the format to be adopted is provided. If not, the instructions to authors' section for a scientific journal can be used. In the case of a pharmacology project, instructions for authors provided by British Journal of Pharmacology are suitable. Starting on the introduction section can provide a challenge to new authors. As always, a clear plan must be made. To compile the contents of the introduction section, readers should be introduced to the topic in stages. It is wise to start with a broad statement about the topic in relation to the broad field of pharmacology. Subsequent sections then lead the reader into the main aspects of the subject. A final section then states the hypothesis or aim of the study in context with what is known to date. As an example a fictitious study entitled 'Inhibition of neutrophil activation by angiotensin receptor antagonists'. A possible Introduction might consist of a series of paragraphs on the following topics.

1. Key role of neutrophils in inflammation.
2. What are neutrophils and how are they involved in inflammation?
3. What are ARBs and what are its known actions?
4. What is known about the interaction of ARBs with neutrophils?
5. Hypothesis: Irbesartan inhibits neutrophil activation by inhibiting lipoxygenase.

The Materials and Methods section can appear deceptively straightforward, but it must conform to the correct format. The purpose of this section is important in that the aim is to provide sufficient information so that the experiments can be reproduced in another laboratory. It should not consist merely of lists, but should be written in sentences.

The materials used should be stated in full, including their source. This includes all chemicals, antibodies and specialist materials. If there are different grades or preparations of a chemical available, the exact one used should be specified. The methods should be described in detail and correctly referenced. Details of any statistical tools that have been should be given.

The Results section should be carefully planned. Results are not necessarily presented in chronological order as a list of experiments, but introduced in a manner that introduces the experimental results in a more logical way. It is usual to present results as a summary of a series of replicate experiments. The form in which they will be presented (tables or graphs) should have been decided earlier in the preliminary presentation stages. Each table and graph must be given a number followed by a fully explanatory legend in a way that explains the results without reference to the text. The methods used should be very briefly mentioned in the legend, together with the results as any statistical tests performed. The results section should not consist of merely a series of graphs and tables, but must have text describing these results with any other details not shown in the illustrations. No conclusions or discussion of the results should occur in this section.

In the first paragraph of the Discussion section it is useful to state the overall advances resulting from the study compared with what was known earlier. Here, deductions or implications from each of the results should be discussed. Again the order in which results are discussed should be carefully planned, reserving a paragraph for each topic or type of experiment. This is the place to compare the results obtained with those previously published. If there are any discrepancies between earlier results, suggestions can be made that may account for these. The concluding paragraphs reiterate the advances made by the study and suggestions made in what direction future studies might take. However, care should be taken to avoid extrapolation or unwarranted speculation.

10.5 PHARMACOLOGICAL LITERATURE

At the present time, it is common for students to start to search for information on a pharmacological topic in a rather random fashion by using a search engine such as Google. One of the first citations that appear will usually be for a Wikipedia citation. It is not unknown for students to attempt to plagiarize this, or spend some time attempting to alter the words without altering the meaning in order to evade plagiarism detection software (see Section 10.6). This is not the best way to start a literature search. It is vital to know how to go about searching for information on a topic and how to evaluate information sources.

There is a hierarchy of reliable and credible scientific information. References that have good credibility are found in peer-reviewed journals, whilst information posted on the internet without any peer review should be viewed with caution. Peer-reviewed articles can be graded as primary sources (reports of experiments directly done by the authors), secondary (a review of a number of primary reports) and tertiary sources (text books and monographs that use information from both primary and secondary sources). Journals themselves form a hierarchy of prestige. The impact factor (IF) is a reflection of the number of citations to the work from other publications, is accepted as a good guide to impact on the field. This can be obtained by checking on the journal's web site, and other journal rating sites. It should be noted that the rating in a list of journals will vary from year to year, so should be checked at intervals. The 'Impact Factor 2012' for a journal would be calculated as follows:

$X =$ the number of times articles published in 2010–2011 were cited in indexed journals during 2012

$Y =$ the number of articles, reviews, proceedings or notes published in 2010–2011

IF $2012 = X/Y$

The highest rated journals are usually the general scientific journals such as Nature (IF $= 36$) and Science (IF $= 31$). Within each field, journals are also graded by IF. Top of the league in pharmacology are titles such as Pharmacology Reviews (IF $= 22$) and Trends in Pharmacological Sciences (IF $= 17$). The highest rated journal publishing primary source papers are Molecular Pharmacology (IF ~ 6) and British Journal of Pharmacology (IF ~ 5). This is a complex subject, as the IF is a mean of all the papers published over the assessment period. It does not reflect the readership or importance in the field of any individual paper.

Information obtained from an internet search engine will yield sources of variable credibility and should be examined to see if it is peer reviewed and contains any primary sources from journals. A problem with search engines is that they do not select information on an unbiased basis. Rather internet sites are selected using reiterative algorithms programmed to select the most frequently retrieved sites, irrespective of their provenance, accuracy or authority. It is notoriously difficult to erase discredited information published in web sites. Wikipedia is also inappropriate for unbiased and accurate academic information, since the author is unknown and information subject to random alterations. An improved search is obtained using sites such as Google Scholar, which does a broad search of academic articles, including books and academic journals. Many of the papers published in highly rated peer-reviewed journals are not retrieved by Google Scholar. It is important to note that if information in a citation obtained from the internet comes from a published journal or book, the actual journal citation should be given, and not the URL from which the information was obtained. This equally applies to e-published journals. The journal, date, volume and page numbers should be cited. Unpublished or unverifiable information is suspect and has little reliability. Frequently it comes from an author who has a commercial bias or interest, or from an author who has some personal agenda.

An accurate and targeted search of academic journals is best done using a dedicated database site such as that run by the National Center for Biotechnology Information (NCBI) at the National Institute of Health (NIH). NCBI runs databases for journals (PubMed) as well as databases for nucleotide, gene and taxonomy sequences. PubMed (www.pubmed.com) is an invaluable site with free access. It gives access to millions of citations for biomedical literature from MEDLINE. Most colleges and universities have an institution-supported database system such as Athens, which allow access to many e-published sources not available elsewhere. Students are advised to contact their library information services to learn about how to use these services. The use of sophisticated literature databases requires some experience. A typical example is that of a student's need to carry out a literature search on a new topic, perhaps for a project or essay. The answer is not to start by looking for in-depth information, but to get a broad survey of the topic. The first place to look is in text books and review journals. The first search using PubMed should be limited to reviews using some carefully selected keywords. In the pharmacology area, a respected and highly rated review journal is the TRENDS series of journals published by Elsevier. Other titles are Trends in Pharmacological Sciences and Trends in Neurosciences, but this series covers all the life sciences. Other highly rated review journals are the Annual Review series, which covers all areas of science. Of particular interest to pharmacologists is Annual Review of Pharmacology and Toxicology. Pharmacology Reviews contains in-depth reviews covering the broad field of pharmacology, as well as updates on nomenclature.

10.6 HOW TO CITE SCIENTIFIC
INFORMATION SOURCES

It is important to appreciate the difference between a reference list and a bibliography. A reference list must contain sources cited in the text only. Sources containing relevant information not cited in the text are placed in a bibliography.

There are two main formats for citing references, the Harvard and Vancouver systems. Instructions stipulating which should be adopted are given by individual journals and the authorities governing the regulations for project reports. The Harvard system is essentially an alphabetical listing of the references, whereas the Vancouver system references are assigned a number and references are listed numerically. There are strict rules about how to cite references, discussed in detail in a useful book for students by Pears and Shields (2010). If these are not adhered to the report may be rejected.

Briefly, using the Harvard System (as in this book), authors are listed in the text by last name only with no initials, followed by the date of publication. If there are one or two authors, they are both cited, but if there are more than two authors, only the reference is cited as the first author followed by *et al.,* and the date. If more than one reference is cited together, then they are separated by a semicolon. Citations are listed in the references section by giving the full names of all the authors (last name followed by initials only). The date is then given, followed by the title of the paper, the volume and the page range. Note that they are not listed in the reference list by number. When using the Vancouver system, each reference is assigned a number and only this number is cited in the text. In the reference list, the references are listed numerically, and the reference is given in full as described for the Harvard system.

References from books are cited in the text as for articles from journals. In the reference list, the authors with their initials are given followed by the date and title of the book. The edition of the book should be given, together with page numbers and the publisher. Sometimes the ISBN number (International Standard Book Number) is also given. Other sources of information are sometimes used, such as newspaper articles, manuals or personal communication. These are of lesser scientific credibility and so should be avoided if possible. These should be cited using as much information as possible so that they could be verified. Citation of web sites is controversial as there appears to be no universally accepted format. Individual journals may stipulate their own required formats. Guides for citing information from the internet are given by many individual universities but a particularly comprehensive guide is given by the Learning Centre of the University of South Wales (2013). Information obtained from the internet can be of many diverse types – web sites, database items, documents within web sites, online articles and e-books. The minimum information that must be given such that a reader can access and verify the information is an accurate and full URL (Uniform Resource Locator or Web address) and access date. Articles published in online journals or cited in databases should be scrutinized to see if there is a paper version available, in which case this should be cited. Articles posted online ahead of publication are given a DOI (Digital Object Identifier), in which case this should be given, together with authors, date and title.

10.7 PLAGIARISM

All students must be aware that plagiarism is a most serious academic crime that is likely to be punished by expulsion. It is widely understood to be copying, it is less widely appreciated that it is a broad term to encompass the use of work as being the original creation of the author. It is usually theft of another person's work, but also covers the use of the author's work that has already been submitted for another assignment. Many students appear to think that merely rewording the wording of an original source is sufficient to evade the charge of plagiarism. It may be possible to reduce the copying score registered by plagiarism-detecting software, such as Turnitin, but it may not escape closer inspection. The term 'plagiarism' encompasses the appropriation as the wording, ideas, thoughts or research strategies of another without due acknowledgement. It is not enough to copy sections of work and then merely include a reference. To avoid the charge of plagiarism it must be clear that the author has organized the work in an original way and original words.

REFERENCES

Piers, R. and Shields, G. (2010) *Cite Them Right. The Essential Referencing Guide.* Palgrave Study Skills, Palgrave Macmillan.

Learning Centre, University of New South Wales. (2013) *Harvard Referencing. Electronic Sources,* http://www.lc.unsw.edu.au/onlib/pdf/elect_ref.pdf (accessed 2 June 2013).

Appendix 1

Molecular Weights of Commonly Used Drugs

MOLECULAR WEIGHTS (MW OR FW) OF COMMONLY USED DRUGS

Drug salt	MW of salt	MW of base	Comment
(Atropine)$_2$ • SO$_4$	676.8	289.4	
5-Hydroxytryptamine • HCl	212.68		Add ascorbate
Acetylcholine • Cl	181.7	146.2	Hydrolyses in dilute solution
Adrenaline bitartrate	333.3	183.2	Add ascorbate
Carbamylcholine • Cl	182.7	147.2	Carbachol
Clorgyline	308.6		
Diphenhydramine sulphate	291.8	255.4	
Diprenyl • HCl	223.7		Selegiline
Adrenaline			See adrenaline
Eserine salicylate	413.5	275.3	
FMLP	437.55		
Hexamethonium bromide	362.2	256.3	
Histamine dihydogen phosphate	307.1	111.2	
Isoprenaline • HCl (isoproterenol)	247.7	211.2	Add ascorbate
Methacholine	195.7	160.2	
Neostigmine • Br	303.2	267.7	

(*Continued*)

Drug salt	MW of salt	MW of base	Comment
Nicotine hydrogen tartrate	462.4	162.2	
Noradrenaline hydrogen tartrate	337.3	169.2	Add ascorbate
Noradrenaline			See noradrenaline
PAF (γ-O-hexadecyl)	523.68		
Phenylephrine • HCl	203.7	167.2	
Phorbol 12-myristate 13-acetate	616.83		PMA
Physostigmine			See eserine
Propranolol • HCl	295.80		
Salbutamol sulphate	288.35		Salbutamol • $\frac{1}{2}H_2SO_4$
Serotonin			See 5-hydroxytryptamine
Succinylcholine			See suxamethonium
Suxamethonium • Cl • $2H_2O$	397.3	290.3	
Tranylcypromine • HCl	169.6		
Tubocurarine	771.72		Hydrochloride • $5H_2O$
Acetly-β-methylcholine • Cl			See methacholine

Appendix 2

Useful Resources for Practical Pharmacology

There are a number of useful resources that are supplied by commercial suppliers of chemicals and equipment that are valuable in planning and executing pharmacological experiments. In most cases, these consist of free downloadable booklets in pdf format.

A particularly useful resource is the Glaxo Pocket Guide to Pharmacology by Sheehan and Elliott, first produced in 1993. Easy access to this is made by inserting "Glaxo pharmacology guide" into a search engine and ensuring that the correct URL has been obtained: http://www.pdg.cnb .uam.es/cursos/Barcelona2002/pages/Farmac/Comput_Lab/Guia_Glaxo/ intro.html, accessed on 24/10/2013. This contains brief definitions of terms used *in vitro* and *in vivo* pharmacology, together with a compact summary of receptor classification systems.

Another resource is supplied by Tocris Bioscience (www.tocris.com, accessed on 24/10/2013). Under Resources, several pdf documents are found, including a glossary of pharmacological terms and a list of common abbreviations. For those who feel less confident in performing calculations involving molarities and dilutions, two calculator applications are included.

Merck-Millipore (incorporating CalBiochem and NovoBiochem) provide a useful booklet on their website (www.merckmillipore.com, accessed on 24/10/2013) entitled LabTools that includes a handy explanation of safety data, including international hazard symbols.

Practical Pharmacology for the Pharmaceutical Sciences, First Edition. D. Michael Salmon.
© 2014 John Wiley & Sons, Ltd. Published 2014 by John Wiley & Sons, Ltd.

Several other sites are useful to check molecular weights of chemicals and their slats, together with numerous protocols. Amongst these are Sigma-Aldrich (www.sigmaaldrich.com), Fisher Scientific (incorporating Thermo Scientific) at www.fisher.co.uk, accessed on 24/10/2013 and Life Technologies (incorporating Invitrogen) at www.lifetechnologies.com.

Two major suppliers of equipment and apparatus are AD Instruments and Harvard Apparatus. AD instruments (www.adinstruments.com) has been frequently referred to in this book as suppliers of hardware and software that are invaluable in performing pharmacology experiments. Their Education section is particularly useful in providing background information and some protocols for physiology and pharmacology experiments using their equipment.

Harvard Apparatus (www.harvardapparatus.com) supply a wide range of specialized equipment for pharmacology experiments, including organ baths and electrodes.

Index

Acetylcholine
  on heart, 94, 103–104
  on ileum, 28–29, 55
Acetylcholinesterase (AChE)
  assay, 142, **143–145**
Acetyl-β-methacholine, see
    Methacholine
Acetylthiocholine, 143–144
Acid citrate dextrose, 127
AD Instruments, 49–50, 58, 68, 99
Adrenaline, 36, 79–81
Adrenoceptor, 30, 36, 56
  aorta, 103
  heart, 97–99, 102
  ileum, 78–80, 82
  vas deferens, 82
Agonists, 27–30
Ahlquist, R.P., 36
Alzheimer's disease, 142
Analysis of variance (ANOVA), 17, 21
Animals (Scientific Procedures) Act
    (1986), and Amendments
    (2012), 5
Antagonism
  mechanisms, 30–37
  selective, 59–62
Antilogarithm (antilog.), 11
Anxiolytic drugs, **163–165**, 171–172
Aorta rings, 102–104
Apocynin, 133
Ascorbic acid, 189–190
Atenolol, 34
  on heart, 102

ATP-ase, assay, 154–158
ATP, cotransmitter, 82
Atria, isolated, see Auricle
Atropine, 30, 34
Auerbach's (sub-mucosal) plexus,
    55–56
Auricle, isolated, 101–102
Autonomic nervous system, 36
Avagadro's number, 8

Beer-Lambert law, 133–134
Benzodiazapines, 163, 175
Boassays, 37
  bracketing, 38–39
  4-point, 39–40
Bonferroni post-test, 17, 21
British Pharmacological Society, ix,
    162, 183
Butryl (pseudo-) cholinesterase, 62
Butyrylcholine, 58–59

Calcium channel blockers, 73–74
Cannulation, 95
Carbachol (carbamoyl choline)
  and cholinesterases, 62, 85, 89, 143
  skeletal muscle, 110, 116
β-carbolines, 30
Cardiac glycosides
  heart, 94
  Na^+, K^+-ATPase, 142, 155, 158
Catecholamines, 36, 101
Categorical or survival data, 12, 22
Chart Reader® software, 52

Practical Pharmacology for the Pharmaceutical Sciences, First Edition. D. Michael Salmon.
© 2014 John Wiley & Sons, Ltd. Published 2014 by John Wiley & Sons, Ltd.

Chart® software
  aorta, 104
  auricle, 99
  heart, 99–100
  ileum, 49
Chemical Hygiene Plan (CHP), 1
Cheng-Prusoff relationship, 34, 154, 158
Chi –squared (χ^2) test, 21–24
Cholinergic nerves, 36
Cholinesterases, specificity blood, 62–63
Cholinesters, 57–59, 62–63, 88
Cholinoceptor, 94, 97, 115, 117–118
Chromogenic p-nitroanilide substrates, 151–152
Chromozym TH, 151–153
Chronic granulomatous disease (CGD), 131
Chronotropic, 94
Cimetidine, 34
Clorgyline, 146–151
Concentration ratio (CR), 32–33
Confidence interval (CI), 16, 20
Confocal microscopy, 134
Contact time, 50–51
Contingency tables, 21–25, 38
Continuous (or quantitative) data, 12
Controls, (positive, negative), 6
Control of Substances Hazardous to Health (COSSH), 1–2
Coulter counter, 122, **124–125**
Cultured cells, 121–122
Cyclic AMP heart, 98
Cyclic GMP, 128, 130
Cytochrome c reductase, 132

Dale, H.H., 36
Degrees of freedom (d.f.), 15, 24
Dependent variable, 6, 17–20
Depolarizing blocker, 109, 117, 119
Deprenyl, 146–151
Desensitization, 60
Diacylglycerol, 134
Dilution, 4, 7, 9
  factor, 9
  ratio, 9
Diphenhydramine, 60. 62, 86–87, 90–92
Diphenylene iodonium (DPI), 133
Discussion section, 184
Disposal, of chemicals, biological material, 2–3

Dithiodinitrobenzoate (DTNB), 143–144
DOI (digital object identifier), 188
Dopamine, 145
Dose-ratio, *see* Concentration ratio, CR
Dosing, 50

Eadie-Hofstee plot, 169–170, 174
EC_{50}, 28–29
EDRF, 103
Efficacy, 29–30
Ehrlich, P., 27
Einthoven, W., 52
Electrically-stimulated preparations, 52–53
  field, 45, 52, 83
  transmural, 52–53
Electrochemical detection, 166
Electrodes, stimulating, 45, **52**
Elevated Plus Maze (Neophobia) test, 163–164
Ellman reaction, 143
Epinephrine, *see* Adrenaline
Equivalent, 8
Eserine (physostigmine), 63–64
  skeletal muscle, **111–116**, 120
European Agency for Health and Safety, 2
Experimental design, 5, 7, **22–23**

Fasciculation, 119
Ferricytochrome c, 132–133
Field stimulation, 45, **52**, 83
Finkleman, B., 52
Fisher's exact test, 21–25
Flow cytometry, 123–125, 134
Fluorescent calcium indicators, 135
Fluoroscan Ascent reader, 136
F-met.leu.phe (FMLP), 131–138
Focal innervation, *see* Skeletal muscle innervation
Frog rectus abdominis, 108–109, 111
Fura2, 134
  calibration, 136–137
  fluorescence spectrum, 135
Fura2-AM, 135–136
Furchgott analysis, 35–36
Furchgott, R.F., 35–36, 103
F value, 17

Gaddum equation, 31
Gaddum, J.H., 31

Galvani, L., 52
Ganglionic nerves
  pre-, 56
  post-, 56, 83
Gastrocnemius muscle-sciatic nerve
    preparation, 117, 119–120
Gastronemius muscle, 108
Gaussian (normal distribution), 12–14
Geller-Seifter conflict tests, 163–164
Genesis® software, 127
G-protein, 30
GraphPad Prism, 12, 15–16, 20–21
Graph plotting, 178–179

Hanes-Woolf plot, 169–170, 175
Harvard reference system, 187
Hazards, biological, radioactive,
    cancerogenic, 2–3
Health & Safety Executive (HSE), 1
Heart, isolated preparations, 93–99.
    See also Langendorff
Hemocytometer, Neuberger, 122–124
HEPES buffered saline, 126–127
Hexamethonium ileum, 60–62, 85, 89
  heart, 94
Hill, A.V., 27
Hill slope, 20
Histamine
  heart, 94
  ileum, 34, 36
  trachea, 81
Histopaque, 132
HL-60 cells, 122
HPLC, 166
5-Hydroxyindole acetic acid, 166–170
4-Hydroxyquinoline, 147–148
5-Hydroxytryptamine (5–HT)
  MAO substrate, 145
  metabolism, 166–170
  trachea, 81

IC_{50}, 129
  MAO inhibitors, 145
IEMS microtitre-plate reader, 127
Impact factor (IF), 185
Independent variable, 17–20
Innervation, focal, 107–9, 116–8
  multiple, 108–9, 111, 117–8
Inositol, 1,4,5- trisphosphate (IP_3,
    $InsP_3$), 141
Inotropic, 94
International Pharmacopoeia, 37
Intracellular [Ca^{2+}], 131–139

Inverse agonist, 30
Inverse logarithm, 11
ISBN (International standard book
    number), 187
Isometric transducer, see Transducer
Isoprenaline
  on heart, 98–102
  on ileum, 79–81
Isoproterenol, see Isoprenaline
Isotonic transducer, see Transducer

K_A, 31–32
K_B, 31
Ketamine, 30
Krebs-Henseleit solution
  aorta, 103–104
  composition, 48
  heart, 95–102
  rabbit duodenum, 79
Kynuramine, 146–150

Laboratory record book, 3–4
Langendorff preparation, 93–102
Langley, J.N., 27
Law of Mass Action, 28
Leech dorsal muscle, 111–113
Leukotriene B_4 (LTB_4), 131–138
Leukotrienes, 81–82
Literature searches, 185–186
L-NAME, 104–105
Locomotor activity (LMA) tests, 164
Loewi, O., 52
Logarithms, 7, **10–11**
Losartan, 14–18, 22

Markwell, see Protein assay
Materials and methods, 183–184
Materia Medica, 27
McEwan solution, 48
Meissner's (myenteric) plexus, 55–56
Methacholine (MCh), 58–62, 85, 89
Micotitre-plate, 121–132
Miniature end-plate potentials
    (MEPPs), 107–108
Molality, 8–9
Molar, 8–10
Molar extinction coefficient, 133–134,
    141
Mole, 8–9
Monoamine Oxidase (MAO)
  assay, 146–148
  inhibitors, 142–145, 149–151
  subcellular distribution, 146–149

Multiple innervations, *see* Skeletal muscle innervation
Multiskan reader, 127
Muscarine, 36
Muscarinic, 30, 34, 60, 65
Myenteric plexus, *see* Meissner's plexus

NADPH oxidase, 131–133
Na^+, K^+-ATPase, 142
  assay, 154–157
National Centre for Biotechnology Information (NCBI), 186
National Institute of Biological Standards and Control (NIBSC), 37–8
Negative control, *see* Controls
Neophobia, **163–164**, 171
Neostigmine, 143–145
Neuberger haemocytometer, *see* Hemocytometer
Neuromuscular junction, 107–108, 111, 116–118
Neutrophils, 122, 131
  intracellular $[Ca^{2+}]$, 134–139
  NADPH oxidase, 132–134
  preparation, 131–133
Nicotine
  guinea-pig ileum, 60–62, 85, 89
  heart, 94
  skeletal muscle, 109–111, 115–116
Nicotinic receptor
  neuronal, 59, 59–60
  skeletal muscle, 94, 107–118, 120
Nifedipine, 74–77
Nitric oxide (NO)
  aorta, 94, 103–104
  platelets, 127–130
Nitric oxide synthase, 104
Nitrosoglutathione (GSNO), 128–130
2-nitro, 5-thiobenzoate (TNB), 143–144
Non-adrenergic, non-cholinergic transmitters (NANC), 82
Non-depolarizing blocker, 109, 117
Noradrenaline, 36
  heart, 94, 102
Norepinephrine, *see* Noradrenaline
Normal (distribution), 12–14
Null hypothesis, **13–16**, 24

Occupational Safety and Health Administration (OSHA), 1
ODQ (1H-[1,2,4]Oxadiazolo[4,3-a]quinoxal-in-1-one), 128–130

Opioid receptors, 83
Organ baths, 50–52, 54
Ouabain, 155–157
Oxidative burst, 131
Oxyhaemoglobin, as NO scavenger, 128

pA_2, 32–34
PAF (platelet-activating factor)
  neutrophils, 131, 133, 138–139
  platelets, 126
pA_h, 35
Parametric tests, 12–13
Parasympathetic innervation, 52
pA_x, 32
pD_2, 28
Pearson coefficient, 18–19
Perfusion, 50, 52
pH, 11
  of Ringer, 37
Phenylephrine, ileum, aorta, 104–105
Phosphatidyl-inositol, 4,5,bisphosphate, 134
Physiological buffers, 44, 46, **48**, 50
Physostigmine, *see* Eserine
pI_{50}, 145
Plagiarism, 188
Platelet activating factor (PAF), 126, 131, 133
Platelet aggregation, 14–16
PMA (phorbol, 12-myristate, 13-acetate), 131–133
Positive control, *see* Controls
Posters, 180–181
Post-tetanic potentiation, 118
Potency, **27–29**, **31–34**, 37, 39
PPACK (D-Phe-Pro-Arg-chloromethylketone), 152–154
Primary cultured cells, 122
Project reports, 183–184
Proof of concept, 7
Propranolol, 79–81
Protein assay, Markwell, 148
PubMed, 186
P value, 16, 21, 24–25

Reactive oxidative species (ROS), 131
Receptor-occupancy, 29–31, 35
Receptor sub-types, **34**, 36–37
Recording equipment, 44, **49–50**
Rectus abdominis muscle, 108–111
Reference citation, 181–182, 185, **187–188**

Regression
  linear, 12, 17
  non-linear, 17–18
Respiratory burst, *see* Oxidative burst
Resting tension, 54
Results section, 184
Ringer solutions, 48, 54
Risk assessment, 1–3

Salbutamol, 36
  heart, 98–101
Saxby electrode, 78–79
Scatchard plot, 169–170, 174
Schild, H.O., 64–66, 75–77, 86, 91
Schild plot, **64–66**, **75–77**, 86, 91
Selegiline, *see* Deprenyl
SI system, 7
SkanIt® software, 127
Skeletal muscle innervation
  focal, 107–109, 116–118
  multiple, 107–112, 117
Skinner box, 163
Slope, 18
Sodium nitroprusside, as NO donor
  aorta, 104
  platelets, 128–130
Square-wave stimulation, 52–53
Standard deviation (SD), 12–13
Standard error of the mean (SEM),
  14
Stimulators, 52
Student's t-test, 14, 17
Succinylcholine, 118

Superfusion, 50
Superoxide dismutase (SOD), 133
Suxamethonium, *see* Succinylcholine

Tachyphylaxis, **60**, 62, 89
Tacrine, 143–145
Tetanus, 117–120
Theophylline, on heart, 98–101
Thrombin, assay, 151–155
Time (dose) cycle, 51
Trachea, 34
Transducer
  isometric, 47–49
  isotonic, 44–45
Tranylcypromine, 146–151
Trypan blue, 122
t-Test, paired, unpaired, 14, 16
T-tubule system, 118
Tubocurarine, 109–110, 114–116
Tyrode's solution, 47–48

URL (Uniform resource locator, web
  address), 188

Vancouver reference system, 187
Variability, 12, 14, 18
Vas deferens, 182–183
Volatile compounds, handling, 3
Volta, A., 52

Wilcoxon signed rank test, 14

Zaprinast, 128–130